Reinforced Plastics:
Theory and Practice

Cahners Practical Plastics Series

REINFORCED PLASTICS

Theory and Practice

Second Edition

M. W. Gaylord

Senior Engineer, Koppers Company, Inc.

Cahners Practical Plastics Series

CAHNERS BOOKS
A Division of Cahners Publishing Company, Inc.

89 Franklin Street, Boston, Massachusetts 02110

Library of Congress Cataloging in Publication Data
Gaylord, M   W   1917–
  Reinforced plastics.
  (Cahners practical plastics series)
  Bibliography: p.
  1. Reinforced plastics. I. Title.
TA455.P55G39   1974   620.1'923   74–9842
ISBN 0–8436–1210–X

The author believes that the information contained herein is accurate and reliable as of the date of this publication but assumes no obligation or liability that may arise as a result of its use. While the author has no knowledge that the information contained herein infringes any valid patent, he assumes no responsibility with respect thereto, and each user must satisfy himself that his intended application, process, or product infringes no patent.

Printed in the United States of America
Halliday Lithograph Corporation, West Hanover, Mass.

Contents

Preface

The first edition of *Reinforced Plastics* was prepared with a specific objective in mind: to assist the engineer, architect, and fabricator and their associates in industry in the intelligent use and application of reinforced plastics (RP).

Experience from the first edition indicates that this book can be of value to others, too. It has been accepted as a text in universities, technical trade schools, and educational organizations. Salesmen of supplying companies have found the information to be helpful.

In general, discussions have been limited to reinforced plastics composed of polyester resin systems reinforced with glass fibers, with the end product generally in the form of a laminate.

It is intended that the reader will be able to use some of the fundamentals presented relating to reinforced plastics (thermosetting types) to assist him in designing products, selecting materials to meet end-use requirements, specifying materials, deciding on appropriate end-use applications, using these materials in maintenance and repair, and taking advantage of the properties of these materials over other conventional materials.

Acknowledgments

The author gratefully acknowledges his indebtedness to the Koppers people who assisted in the preparation of this book, especially Dr. William T. Booth of the Organic Materials Division, R. Sherman Detrick of the Research Department, and B. Otto Wheeley, senior vice-president.

M. W. GAYLORD

Pittsburgh, Pennsylvania
1974

Part One

THEORY

What Are Reinforced Plastics?

For purposes of this text, the term *reinforced plastics* refers to a directed combination of a thermosetting resin and a reinforcement to produce a composite material, or laminate. Unless otherwise indicated, the resin systems discussed will be unsaturated (reactive) polyester in combination with reinforcements made from glass fibers. Obviously epoxy resins or other thermosetting resin systems which cure chemically or harden after the addition of a catalyst or curing agent, as well as other types of reinforcements, can be used to produce reinforced plastic laminates. The term *laminate* is meant to include all types of reinforced plastic shapes such as solid rods and bars, tube and pipe, structural shapes such as angles, channels, and I-beams, flat sheets as well as housings or any unusual contoured shapes to meet a particular end-use application. These laminates may be produced by hand lay-up methods, machine methods or a combination of both. The term *structure,* in general, refers to a finished article fabricated from one or more laminates. The various processes used to produce laminates in industry are discussed in chapter 4.

The designer or manufacturer can design or produce a reinforced plastic laminate whose mechanical properties in any given direction are both predictable and controllable. This can be done by carefully selecting the resin system to be used in terms of its desired end-use properties, and the reinforcement, in terms of its composition and orientation in the finished product. Where laminates are made by machine methods—particularly filament winding and continuous pultrusion—where the orientation of the reinforcement is carefully controlled, reproducibility of properties in the laminates is greatly increased when compared to laminates produced by hand methods.

By working with these composite materials, resin and fiber glass, the designer or manufacturer is not limited by the physical and mechanical properties of conventional natural materials.

In the short history of their commercial application (since about 1946) unfounded claims frequently have been made for reinforced plastics, as they have been made for other materials. Experience teaches that each and every material has its own peculiar properties and its own specific applications. Even though reinforced plastics have been successfully applied to a wide variety of end uses, this does not justify their indiscriminate use to replace other materials. Reinforced plastics have been misapplied and it would be less than forthright to state that they have no limitations.

The guidelines below may help in deciding whether glass-reinforced plastics are suitable materials for a particular end-use application.

ADVANTAGES

1. Large complex shapes, repairs, or replacements can be readily produced or fabricated.
2. Laminates made by machine processes exhibit reproducible physical, mechanical, and electrical properties.
3. Greater freedom of design is possible with glass-reinforced plastics than with most other materials.
4. Highest strength-to-weight ratios are offered.
5. They are extremely resilient and tough and will not dent like metal.
6. In addition to possessing good weathering properties, they do not readily corrode. They are resistant to attack by many chemicals as well as to mold and fungus attack.
7. Glass-reinforced plastics exhibit good electrical insulation properties as well as thermal insulation properties.

DISADVANTAGES

1. Cannot be recommended where relatively high-temperature (above 400°F) environments are involved.
2. Rigidity is not very high when compared with that of some metals.
3. Properties of handmade laminates may be more difficult to reproduce repetitively.
4. Cost of raw materials is relatively high but may be offset in terms of installed cost, or because of less costly equipment or less highly skilled labor.
5. Some laminating processes may be slow by comparison to those used for metals.

Reinforced plastics should be seriously considered when the following factors are involved:

1. Environmental conditions indicate that conventional materials will not provide expected service life.

2. Electrical insulation or thermal insulation properties are desirable.
3. Impact damage is possible or probable and repairs by conventional methods are restricted. In reinforced plastics, repairs may be effected by simply mating the fractured edges and bonding or patching.
4. Weights of replacement parts will cause extraordinary expenditures in installation costs because of clearance, inaccessibility, or operational downtime.
5. Weight and/or strength retention are problems.
6. Color is required. With reinforced plastics pigments may be introduced into the resin system to provide desired color and eliminate painting costs. Where fluid flow or liquid level height is to be supervised, the translucent laminate will facilitate visual control.
7. Complex shapes or intricate detail present serious problems in working with more conventional materials.
8. Frequent design changes may be expected.
9. Tooling and equipment costs for a competitive material would be unusually high or present delays because of delivery.
10. A single molding in reinforced plastics may replace a complete assembly in a competitive material.
11. The expected production quantity is limited (up to 50,000 units).

Obviously, some of the above criteria apply to a manufacturing or production operation and are not of import to the potential everyday user of reinforced plastics, not engaged in a repetitive type operation.

In the case of the everyday user, what are the features to consider in evaluating an application for reinforced plastics? Simply stated they are:

1. Ease of producing simple and complex shapes as well as fabricating simple or complex structures.
2. An expeditious means of repair compared to replacement of the original material.
3. Lightweight.
4. Simple installation.
5. Machinability.
6. High strength-to-weight ratio.
7. Excellent cost-performance ratio.
8. Minimum heat transfer or heat loss (insulating effect).
9. Nonconductivity (electrically insulating).
10. Little or no maintenance after installation.
11. Noncontamination of product or process.
12. Long service life.

2

Acceptance

Acceptance of reinforced plastics is a matter of public record in the form of demonstrated production figures. Since 1946, reinforced plastics have invaded nearly every area of man's life, from recreational and sporting goods, transportation, building construction, and the defense of his country to the exploration of outer space. The reinforced plastics industry without a doubt has benefited tremendously by the research and development effort sponsored by the defense and space organizations of the United States. Conversely, reinforced plastics have made their contribution to these noncommercial applications.

The reasons for the wide acceptance of reinforced plastics by industry include: 1. Broad chemical resistance; 2. Good mechanical properties; 3. Availability of design criteria to assure equipment reliability; 4. A descending cost trend; 5. Excellent cost-performance ratios; 6. Histories of installations in service for a number of years performing successfully.

What are some of the applications that support this use and acceptance?

Twenty-five years of performance and experience have made the pleasure boat one of the best accepted applications of glass-reinforced plastics for the consumer.

MFG Fishin' Caprice, 18'-2½'' outboard boat of glass reinforced polyester. (Molded Fiber Glass Boat Company, Union City, Pa.)

Corvette, America's only true production sports car, celebrates its 22nd year on the highway. New styling changes its looks and adds to its roadability and durability in 1974. Its reinforced fiber glass body continues to be one of its distinguishing features since it was introduced in 1953. (Chevrolet Motor Division, General Motors Corporation)

RECREATION

In the field of recreational activities the directional strength properties of reinforced plastics are used to advantage in such applications as fishing rods, archery bows, golf club shafts, and lacrosse sticks. Their structural properties and their light weight are evident in their use for canoes, golf carts, snowmobiles, motor homes, vaulting poles, water and snow skis, and surfboards. Other applications include protective helmets (racing and baseball), diving boards and other swimming pool equipment, and playground structures.

Using polyester resins supplied by Koppers, American Insulator Corporation produces lacrosse stick handles. Iroquois Indians on the Tuscarora reservation in New York State fit out the sticks and make them ready for play.

Doubly equipped for either water or snow sports with pressure molded fiber glass reinforced polyester skis and fiber glass toboggans, also usable as a water sled. Lightweight equipment by Kimball-Schmidt, Inc., San Rafael, Ca., using a Koppers Company specially formulated polyester resin.

Sears' twelve-foot Gamefisher in glass-reinforced plastic, produced by Molded Fiber Glass Boat Company, Union City, Pa., is easily handled by two people from cartop to launch. (Owens-Corning Fiberglas Corporation)

(Air Products and Chemicals, Inc.)

Old Town's two man pack canoe, is designed for two people afloat and an easy portage for one. (Old Town Canoe Company, Old Town, Me.)

Strong, lightweight reinforced plastic surfboards have contributed to the rapid growth of a sport which was once enjoyed only by Hawaiian kings. (Camerique photo)

The Xplorer 307 Motor Home features extensive use of glass reinforced plastics throughout its construction. (Xplorer Motor Home Div., Frank Industries, Inc., Brown City, Mi.)

ELECTRICAL APPLICATIONS

The good electrical insulation properties of reinforced plastics are demonstrated in their use for electrical and electronic housings, printed circuit boards, and "hardware" for the electrical utility industry, including guy strain insulators, switch control rods, hot sticks, and switchgear components. Other applications include ladders, "man buckets," and "cherry picker" arms and elbows. Cable trays for use in corrosive environments have proven their worth.

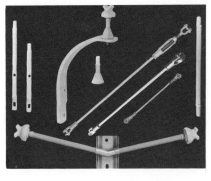

Reinforced plastic components and pole line hardware have contributed greatly to the programs of "beautility" and safety in the electric utility industry.
In addition to their excellent dielectric properties, reinforced plastics provide the necessary strengths with reduced silhouette and weight. The corrosion-resistant, smooth, hard surfaces also resist the imbedment of contaminants.
Utility companies in greater numbers are taking advantage of these components to provide even better service at lower overall cost.

Safety first. 7500 series fiber glass extension ladder manufactured by R.D. Werner Co., Inc., Greenville, Pa. This ladder fabricated with polyester glass reinforced structurals is strong, lightweight, corrosion resistant, and will not rot or warp. It retains its excellent electrical insulating value because it doesn't absorb moisture.

Applications of this "man bucket" include elevated electrical work and tree trimming.

Improved appearance with reduced maintenance costs is demonstrated with the use of reinforced plastic switch control rods.

Examples of electrical apparatus showing the versatility and cost-performance benefits of glass reinforced plastics. Electrical manufacturers cite design freedom, increased reliability, and reduced assembly costs as chief reasons for considering FRP in new and existing applications. (Owens-Corning Fiberglas Corporation)

Men work bare-handed on 138,000-volt power transmission lines, supported and protected by this fiber glass reinforced plastic boom and bucket. (Holan Corp. and Owens-Corning Fiberglas Corp.)

Insulating rail joint for signal system. A densified wood laminate reinforced with unidirectional glass fibers. (Permali, Inc., Mount Pleasant, Pa.)

THERMAL INSULATIONS

Thermal insulation properties make reinforced plastics ideal for motor transport, refrigerator railroad cars, and all-weather shelters.

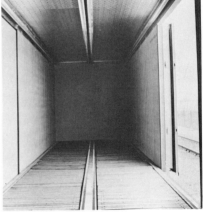

Fiber glass reinforced plastic all-weather shelter. One-piece molded construction fabricated by Warminster Fiberglass Company, a subsidiary of Fischer & Porter Company, Southampton, Pa.

Glass-reinforced plastic housing unit for cryogenic liquid semi-trailer.

The interior walls of refrigerated railroad cars and trucks are now fitted with fiber glass-reinforced sheet, engineered with a selected wave pattern to enhance air circulation and give increased strength and rigidity. These liner panels have revolutionized refrigerated transport . . . made cars more efficient, economical, and much easier to clean.

COMPLEX SHAPES

Reinforced plastics can be formed into complex shapes and irregular contours as exhibited in office and commercial seating, bathroom fixtures, bowling alley ball returns, and automotive assemblies.

Fiber glass-reinforced truck assemblies have solved big problems for both manufacturers and fleet operators. Manufacturers like the idea of molding a complete hood and fender assembly out of reinforced plastic because they save money on tooling and die costs. Fleet operators like the assemblies because they are corrosion-resistant and virtually eliminate maintenance. Ease of damage repair and increased payloads due to reduced weight are additional benefits.

Crucible Products Corporation, New York City, used fiber glass reinforced plastic to produce 5,000 chairs for Kennedy Airport. (Owens-Corning Fiberglass Corporation)

Maximum durability, clean functional design, low-cost maintenance, and molded-in color are combined in this Brunswick Corp. bowling ball return. (Owens-Corning Fiberglas Corporation)

(Fiberform, Inc.)

New MFG Phone Panel for Bell System.
Compression molded from fiber glass
reinforced plastic. No painting, the color
is molded in. It's scratch resistant, with
excellent durability and lightweight, plus
eye-appealing beauty. The book shelf is
stainless steel. (Molded Fiber Glass Company, Ashtabula, Ohio)

The park benches at the New York World's
Fair were among the first to be made from
slats of glass-reinforced plastic.

Stylish molded furniture is strong and
comfortable and virtually maintenance
free. (Owens-Corning Fiberglas Corporation)

Fiber glass bathtub (shower) by U/R Division, Universal-Rundle Corporation, New Castle, Pa.

Vacuum formed acrylic vanity counter, bathtub and sink reinforced with fiber glass polyester resin. Produced by Vacuum Formed Products Division of Volplex Corporation, Rochester, N.Y.

CORROSION RESISTANCE

Corrosion-resistant piping, fuel and chemical storage tanks, scrubbing towers, fume-carrying ductwork, and collection hoods and chemical processing equipment are typical of applications that have been performing satisfactorily for many years. Other successful applications include fertilizer hoppers, grating, cooling tower grids, and equipment used in the electroplating industry.

International Harvester Company Model 400 Cyclo Planter fitted with four 550 pound capacity fertilizer tanks and one 11 bushel seed hopper. Tanks and hopper fabricated from glass reinforced plastics by Molded Fiber Glass Company, Ashtabula, Ohio.

Fiberglas ® gasoline storage tank with 10,000 gallon capacity being lowered into place. (Owens-Corning Fiberglas Corporation)

Two filament-wound fiber glass reinforced plastic tanks used for corrosive chemical storage. (Justin Enterprises, Inc., Fairfield, Ohio)

Water supply lines of glass reinforced plastic installed in a coal mine. (Fiberglass Resources Corporation, Farmingdale, N.Y.)

Chemical process equipment fabricated by The Ceilcote Company, Inc., Berea, Ohio.

Shipment of 20-inch diameter polyester RP pipe.

250 feet of RP pipe can be handled easily by one man.

48-inch diameter pipe, 80 feet long.

A lightweight septic tank made of glass reinforced polyester can be handled easily by two men. The 550 gallon capacity tank weighs only 150 pounds, less than half the weight of steel tanks of the same capacity and 24 times lighter than comparably sized concrete tanks. Other suggested uses are water storage, farm or industrial product storage, underground utility housings, storage of industrial pollutants or chemical slurries. (Rockwell International, Reinforced Plastic Operation, Linesville, Pa.)

Co-current and counter-current packed scrubbers fabricated from glass reinforced polyester by Hastings Reinforced Plastics, Inc., Hastings, Mi.

Nested Parshall flume, precision molded of glass reinforced polyester resin by Warminster Fiberglass Company, Southampton, Pa. Nested construction permits removal of inner flume when additional flow-metering capability is needed.

Firmaline ® Fiberglass reinforced plastic grating produced by Joseph T. Ryerson and Son, Inc., incorporates built-in safety for many industrial applications. It is used in chemical and electroplating plants for its corrosion resistance, on off-shore drilling platforms for its lightweight and weathering properties and by electric utility companies for its electrical insulation properties.

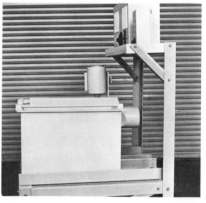

Installation of preinsulated, filament-wound fiber glass reinforced plastic duct runs along the entire 585-foot length of the roof at Kimberly-Clark plant, Beech Island, S.C. (PPG Industries)

Etching and plating racks fabricated from structural shapes.

Chemical Process Equipment fabricated by
The Ceilcote Company, Inc., Berea, Ohio.

AEROSPACE AND MILITARY

The high strength-to-weight ratios of reinforced plastics have been demonstrated in aerospace and military applications in rocket engine cases, nozzles, nose cones, submarine fairings, pressure bottles, wing tip tanks, helicopter rotor blades, and rifle stocks. Additional applications include radomes, assault boats, aircraft lounge seats, engine cowlings, ductwork, and complete aircraft fuselage surfaces.

Towers, with full-scale mock-up of Apollo earth-to-moon manned spacecraft at the NASA Space-Craft Centre, Houston, Texas, for the testing of communication and radar equipment to be used on these flights. Towers were fabricated by Scientific-Atlanta, Inc. of Atlanta, Ga., from reinforced plastic shapes. (World Book Encyclopedia Science Service, Inc.)

A 150-foot diameter metal-space-frame radome. The triangles, nine to fifteen feet on a side, are each covered by a vacuum bag molded membrane of 1/32-inch fire retardant glass reinforced polyester. (MIT Lincoln Laboratory)

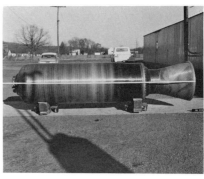

Cherokee 180 is background for a sampling of many glass fiber reinforced plastic parts made by Piper Aircraft Corp., Lock Haven, Pa.

Booster rocket motor filament wound from fiber glass and epoxy resin. (Hercules Powder Corporation)

U.S. Air Force C–9A "Nightingale" hospital aircraft built by McDonnell Douglas Corporation utilizes fire-retardant glass reinforced polyester flooring units to provide tough, lightweight, slip-proof surfaces. (M.C. Gill Corporation)

Unitized 60-foot radome fabricated from reinforced plastic. (Morrison Molded Fiber Glass Co., Bristol, Va.)

BUILDING MATERIALS

Weather resistance is an accepted property of reinforced plastic decorative panels, sliding, and curtain wall materials for residential and commercial buildings. Other successful applications include patio roofs, glazing and skylighting panels, and protective housing for sensitive transcontinental communication systems. In the construction industry concrete pans are in everyday use.

Exterior facia panels fabricated of weather-
and fire-resistant Hetron ® polyester resin
and glass fiber reinforcement. Panels,
fabricated by McClarin Plastics, Inc., of
Hanover, Pa. Building is operations center
for the New Holland, Pa., division of Sperry
Rand Corporation. (Hooker Chemical
Corporation)

Sports arena with base diameter of 206 feet
at Newcastle-upon-Tyne incorporates a
glass reinforced plastic roof built from
Deeglas glass fibre. (Deeglas Fibres Ltd.
and Artrite Resins Ltd.)

FRP concrete forms produced by Molded
Fiber Glass Concrete Forms Company,
Ashtabula, Ohio, in place ready for the
placement of concrete.

Four FRP and foam sections, bonded and
bolted together, mounted on a steel frame,
form the 8 X 18 X 9½ feet structure to
house sensitive television transmission
equipment. The units are produced by
Polyfiber, Ltd., Renfrew, Ontario.

Installation of a 500-lb. RP spire was facilitated by airlifting it to the roof top. Installation took less than 15 minutes, and cost of the installation was cut in half.

A swimclub convertible enclosure for swimming pools. The enclosure consists of a sandwich panel produced by permanently bonding two glass-reinforced acrylic/polyester sheets to a grid core. (Structures Unlimited, Manchester, N.H.)

Glass reinforced plastic sandwich panels, 14' 6" X 6' fabricated by Polyfiber Ltd., Renfrew, Canada for high school near Arctic circle. Panels, constructed of fire-retardant Hetron ® polyester, fiber glass, and urethane core for added insulation, will protect Eskimo teenagers from the region's severe weather. (Durez Division, Hooker Chemical Corporation)

Tub-shower unit of glass reinforced plastic being positioned in family housing complex. (Owens-Corning Fiberglas Corporation)

In other major markets for glass-reinforced plastics some of the accepted end use applications are:

Transportation—Seating, farm equipment, and automotive components
Marine—Ship hulls, ventilation cowls, water tanks, and marker buoys
Appliance—Water-softener tanks, air-conditioning components, laundry tubs, and bathroom fixtures
Protective Covers—outboard engine covers, lawn mowers, and machinery guards

The applications listed here represent only a few of the reported 30,000 or more accepted end uses for glass-reinforced plastics.

Chris Craft 38-foot cruiser. (Owens-Corning Fiberglas Corporation)

Blower fabricated from glass reinforced polyester by The Ceilcote Company, Inc., Berea, Ohio.

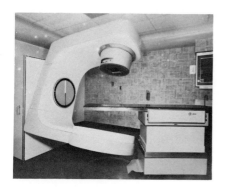

Housing for x-ray unit from glass reinforced polyester. Fabricated by Warminster Fiberglass Company.

Fiber glass roof and innerliner help absorb and dampen sound on these John Deere "Sound-Idea" farm tractors. The components are color molded green and black from sheet molding compound by Goodyear Aerospace Corporation at Jackson, Ohio, to fit the John Deere color scheme.

"Appearance hood" for the outdoor section of a General Electric Executive air conditioner, produced from glass reinforced plastic by the commercial plastics division of Goodyear Aerospace Corporation, Jackson, Ohio.

(Owens-Corning Fiberglas Corporation)

Seamless one-piece Tub-shower molding of gel-coated fiber glass reinforced polyester resin with integral polyurethane foamcore structural ribs produced by Seneca Manufacturing Corp., Zelienople, Pa.

Bright white body on International Harvester's Cadet 75 riding mower is molded into the glass reinforced body by Goodyear Aerospace's Commercial Plastics Division. The fiber glass body made from sheet molding compound resists denting and cannot rust; the integral white color eliminates the problem of peeling paint.

Chrysler Corporation outboard motors feature polyester glass reinforced plastic engine covers to help muffle sound. Goodyear Aerospace's Commercial Plastics Division at Jackson, Ohio, utilizes sheet molding compound on a 200 ton press to produce the 5½ pound hood.

Rustproof plastics, made by matched-die molding glass reinforced polyester, are used for the fenders and instrument console cover (right of driver's seat) on this Ford tractor. The components are made by Goodyear Aerospace Corporation's Commercial Plastics Division at Jackson, Ohio.

Instrument panel for International Harvester Corporation truck produced from sheet molding compound by Goodyear Aerospace Corporation.

The Hatteras 74, the U.S. fishing industry's first all-Fiberglas ® reinforced plastic production line shrimp trawler is manufactured by the Hatteras Yacht Division of Rockwell International. Industry experts say FRP makes possible a stronger trawler that requires little maintenance. The all-FRP fish hold is highly sanitary and easy to clean. (Owens-Corning Fiberglas Corporation)

The list of accepted applications includes items used from the "cradle to the grave," each of which has contributed to the industry's growth of over 3000 percent during the period from 1953 to 1973.

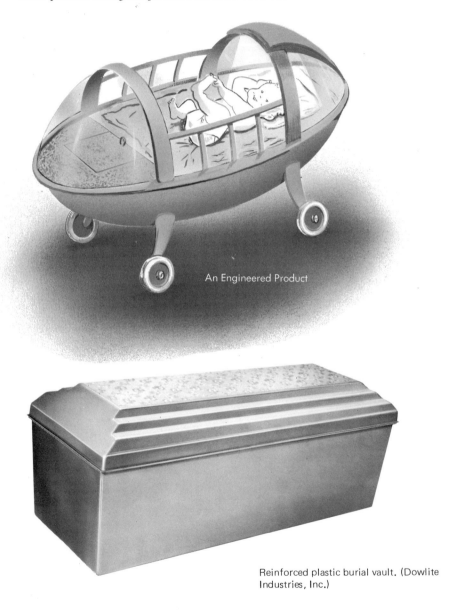

An Engineered Product

Reinforced plastic burial vault. (Dowlite Industries, Inc.)

3

Corrosion Considerations

Corrosive liquids and fumes are deteriorating plant equipment at an annual rate in excess of 10 billion dollars. This annual loss is representative of equipment only and does not include the sizable investments by material suppliers, fabricators, and users to develop corrosion-resistant materials. Added to this dollar loss, the time lost because of plant shutdown to repair corrosion damage is a dollar loss that industry can ill afford. Industry is learning that the high cost due to corrosion in manufacturing can be cut significantly through the use of well-designed and well-applied reinforced plastics.

There are a number of factors that have a marked influence on the service life of reinforced plastic equipment to be used in corrosion service, such as:

1. The type of resin system.
2. The type of reinforcement.
3. The sequence of fabrication of layers.
4. The controlled distribution of resin and reinforcement within the laminate.
5. The proper design of the laminate to meet the stress requirements of the structure.
6. Well-controlled fabrication techniques to assure adequate cure of the resin system and minimize faults such as voids and pinholes.

The importance of fabrication technique cannot be overstressed. In an appropriate application a well-prepared laminate utilizing the proper materials will practically guarantee satisfactory performance. Laminates or structures containing the ideal resin and reinforcement combination but of poor quality workmanship will generally fall far short of expectations.

Experience has shown that the corrosion resistance of a reinforced plastic product is primarily dependent on the construction of the laminate which comes in contact with the corrosive material. This should be a resin layer ranging from 10 to 15 mils in thickness.

Structures or components properly fabricated, either by hand methods or by machine methods (filament winding), can exhibit good corrosion resistance. It is important that the potential customer select a manufacturer who can design and build a laminate that is properly engineered for the particular end-use application.

Approximately 10 million pounds of reinforced plastics were used in corrosion-resistant industrial equipment in 1960. This amount of usage was doubled by 1962–63—a clear indication that reinforced plastics are out of the pioneering stage.

The experienced producer of laminates should take into consideration all of the factors that will affect performance of the final product:

1. Strength and other physical properties required under normal conditions, both static and dynamic.
2. Range of temperature (continuous duty and cyclic).
3. Types and concentration of corrosive solutions or fumes.
4. Presence of possible abusive environment—mechanical or human.
5. Pressure or vacuum to which the equipment will be subjected (continuous duty and cyclic).
6. Design factors necessary for guaranteed performance of a particular piece of equipment.

An additional caution should be considered by the fabricator-vendor. Many well-designed reinforced plastic structures or components have been damaged in shipment to the job or by improper and careless handling at the job site. Any damage or deterioration because of handling or shipping leads to premature attack by corrodents and early failure. These factors can be minimized by proper crating and support and by supervision and proper handling at the point of use.

CORROSION MECHANISMS

The types of corrosion in reinforced plastics are limited and may be classed as:

1. Swelling.
2. Destruction of the resin phase by chemical attack such as:
    a. hydrolysis
    b. oxidation
    c. pyrolysis
3. Loss of adhesion between resin and reinforcement.

4. Destruction of reinforcement—in the case of glass fibers by strong alkali or hydrofluoric acid.
5. A combination of 1, 2, 3, and 4.

Since reinforced plastics are heterogeneous in composition, corrosive attack may be confined to:

1. The resin phase.
2. The interfacial area between the resin and the reinforcement.
3. The reinforcement.
4. A combination of the above.

Metals corrode in an entirely different fashion. Of the 14 types of corrosion in metals and alloys listed in the literature, some of the more common types are:

1. General corrosion—the thickness of metal is gradually reduced.
2. Galvanic.
3. Aerobic.
4. Pitting.
5. Dezincification.
6. Graphitic and intergranular corrosion.

The difference between the effects of corrosive agents on metals and reinforced plastics is important to those who specify equipment. If properly chosen for the service involved, reinforced plastics properly made will last almost indefinitely. After a slight loss of strength during the initial period of use, reinforced plastics can be expected to remain stable and show little further deterioration of properties. Metals, on the other hand, will continue to react with the chemical environment and will continue to deteriorate over a period of time.

Selection of reinforced plastics as well as other construction materials should be based on the installed, long-term cost-performance relationship and not on material cost comparisons alone.

COST-PERFORMANCE RELATIONSHIPS

The following table presents cost comparisons of material only for a 3" x 3" x ¼" angle of four different materials of construction for use in a nonload-bearing (space-filling) application. This size section has been selected since the exposed area is approximately one square foot.

Based on the above comparisons, it is obvious that carbon steel would be the most economic material.

	Reinforced Plastic	Carbon Steel	Stainless Steel 304	Aluminum 6061
Weight, lbs. per lin. ft.	0.94	4.9	4.9	1.68
Price per pound	$1.00	$0.09	$0.70	$0.40
Cost position, per foot	$0.94	$0.44	$3.43	$0.67

Assuming that a mild corrosive environment, one that is exposed to industrial atmospheres or coastal environment, is involved for these materials and that a satisfactory protective coating system must be applied initially and maintained over a period of years, the economic comparisons (total cost position) change significantly over a ten-year period of installed service.

	Reinforced Plastic	Carbon Steel	Stainless Steel 304	Aluminum 6061
Cost Position (see above)	$0.94	$0.44	$3.43	$0.67
Initial cost of protective system	–	0.12	–	–
Repainting costs	–	.45[a]	–	0.45[b]
Total Cost Position	$0.94	$1.01	$3.43	$1.12

a. Estimated cost of repainting during a ten-year period.

b. Estimated cost of maintenance of aluminum (6061) over ten-year period on the basis that an initial protective coating is not required.

Considering the same materials in a severe corrosive environment (industrial chemical fumes, splash, or immersion) the comparison points more favorably toward the use of reinforced plastics. In the case of protective coatings a more sophisticated system would be required to adequately protect carbon steel.

It is obvious that the cost position of a material can change substantially when compared on the long-term, cost-performance relationship.

	Reinforced Plastic	Carbon Steel	Stainless Steel 304	Aluminum 6061
Cost Position (see above)	$0.94	$0.44	$3.43	a
Cost of initial protective system	–	0.40[b]	–	–
Maintenance of coating system	–	1.80[c]	–	–
Total Cost Position	$0.94	$2.64	$3.43[d]	

a. Aluminum would generally not be considered for the severe corrosive environment.

b. Includes cost of sand blasting and application of coating.

c. Includes replacement of coating three times.

d. The cost position for stainless steel might exceed this amount since a more corrosion-resistant type might be required.

The cost comparisons are even more favorable to a corrosion-resistant material if one includes other less definitive costs, such as cost of inspection and supervision and preparation of specifications when protective coating systems are involved. Other costs that are sometimes difficult to define are the increased fabrication and installation costs related to types of construction material that are generally more costly to fabricate or install because of weight factors. No credit or change in cost position has been taken in these comparisons for loss of esthetics that are bound to occur during the period (3 years) between the reapplication of the protective coating systems.

It is possible to develop similar comparisons for materials that would be utilized in structural (load-bearing) applications. In corrosive environments these comparisons would again favorably point to the use of corrosion-resistant materials.

The person selecting equipment should also consider that both surfaces may be exposed to corrosive environments. Although the environments may be substantially different, the effect of corrosion must still be considered because it will affect the installed cost and the expected service life.

There is no question that reinforced plastics will continue to grow in use as a construction material for many applications and as a substitute for conventional materials that are not now providing expected service life.

4

Processes Used in Producing Reinforced Plastics

Many processes or fabrication techniques have been developed for the production of reinforced plastics. Each process has its own usefulness for combining different kinds and amounts of resins and reinforcements (generally glass fibers).

The person interested in knowing more about reinforced plastics should be interested in an overview of all the processes available. This review will provide an appreciation of the amount of technology (whether it be an "art" or a "science") that is involved:

1. Hand lay-up or contact molding.
2. Spray-up.
3. Encapsulation.
4. Filament winding.
5. Centrifugal casting.
6. Continuous pultrusion.
7. Matched-die molding.
8. Continuous laminating.

Detailed discussion of each process is beyond the scope of this book. Some detail has been included on hand lay-up techniques since this process can be useful to those desiring to fabricate structures from reinforced plastics or to make repairs to plant equipment. Little or no capital investment is required in the hand lay-up process. Those interested in pursuing selected processes in greater detail should refer to recognized publications (see Appendix).

HAND LAY-UP

The oldest and simplest process for making reinforced plastic laminates, hand lay-up can be used for epoxies as well as polyesters. A flat surface, a cavity (female) or a positive (male) shaped mold, made from wood, metal, plastics, reinforced plastics, or combinations of these materials, may be used. Fiber glass reinforcements and resins are placed manually against the surface of the mold, and brushes, squeegees, and rollers are used to work the resin into, and to remove air from, the layered reinforcement. Thickness is controlled by the layers of materials placed against the mold. Gel coats may be used as the first layer to provide special surface effects or to form a corrosion-resistant layer of resin.

Variations of hand lay-up include using heat to accelerate the cure, or using vacuum bag, pressure bag, or autoclave techniques to permit the applica-

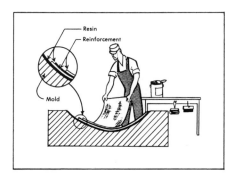

CONTACT MOLDING (Hand lay-up)

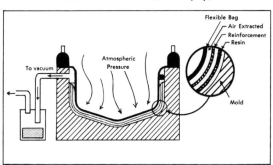

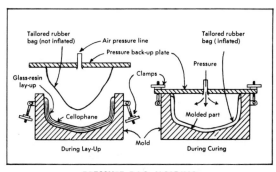

PRESSURE BAG MOLDING **AUTOCLAVE**

tion of pressure against the laminate and force out air (to reduce voids) and excess resin (to increase the glass-to-resin ratio). Pressure techniques influence the final thickness of the laminate as well as the glass-to-resin ratio.

SPRAY-UP

Chopped glass fiber and resin are simultaneously deposited in a mold (generally female). The fiber glass rovings are fed through a chopper and ejected into a resin stream which is directed at the mold by either one of two spray systems: (1) a spray gun ejects resin premixed with catalyst or catalyst alone, while another gun ejects resins premixed with accelerator; or (2) resin ingredients are fed into a single gun mixing chamber ahead of the spray nozzle. By either method the resin mix precoats the strands of glass and the merged spray is directed into the mold by the operator.

After the glass-resin mix is deposited, it is rolled by hand methods to remove the air, compact the fibers and smooth the surface. Curing techniques are similar to those used in hand lay-up and the same techniques to apply pressure during cure may also be used.

SPRAY-UP   ENCAPSULATION

ENCAPSULATION

Milled fibers of glass or chopped strands (short lengths) are thoroughly mixed with the catalyzed resin system and poured into open molds to surround and envelop components, generally for electrical applications. Fibers are used to reduce crazing. Selected fillers may be substituted for glass fibers and are used to decrease shrinkage and increase the useful temperature range of the resin system. Cure may be at room temperature. A postcure at elevated temperature for a predetermined time period may also be desirable.

FILAMENT WINDING

This process, applicable only to surfaces of revolution, uses reinforcements in continuous strands to achieve efficient utilization of glass fiber strength. Roving or single strands are fed from a multiplicity of creels through a resin bath and wound on a suitably designed mandrel. Dry winding using preimpregnated roving may also be employed. Specially designed winding machines similar to lathes orient and lay down the glass in a predetermined pattern to give maximum strength properties in the direction desired. Thickness is dependent on the number of complete layers (passes) uniformly placed on the mandrel. After it has reached the desired thickness, the wound mandrel is usually cured in an oven rather than at room temperature; the mandrel is removed from the laminate after the cure operation is completed. Soluble mandrels may be used to permit winding of closed-end structures.

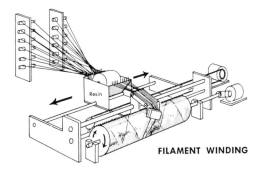

FILAMENT WINDING

CENTRIFUGAL CASTING

Cylindrical structures such as tubing can be formed by this process. Chopped strand mat or fabric is positioned inside a hollow mandrel which can be rotated. Resin mix is added to the rotating mandrel and is distributed uniformly throughout the glass reinforcement by centrifugal force. External heat may be applied through the walls of the mandrel, or the rotating mandrel may be enclosed in an oven. This process is not particularly amenable to room temperature cures, since the equipment may be utilized more efficiently at controlled temperatures.

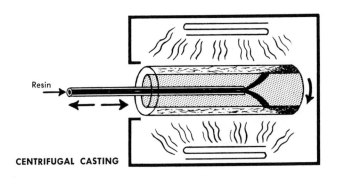

CENTRIFUGAL CASTING

CONTINUOUS PULTRUSION

Selected continuous reinforcements are fed from a multiplicity of creels through a resin bath and then drawn (pulled) through a die which determines the cross-section geometry of the shape and effects control of the resin content. Final cure is accomplished in a heated section of the die through which the shape is drawn by a suitable pulling device.

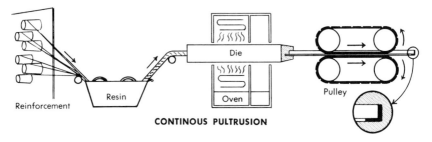

CONTINOUS PULTRUSION

Production of polyester glass reinforced plastic flat sheet by continuous pultrusion. (Morrison Molded Fiber Glass Company, Bristol, Va.)

MATCHED-DIE MOLDING

This process is used for mass production of reinforced plastics. Mat, fabric, or preformed reinforcement is combined with the resin mix at the press just prior to, or just after, placement in the mold. Sheet-molding compound, bulk-molding compound, or prepreg can also be used as a starting material. Heated metal molds form and cure the part at pressures up to, or more than, 300 psi. Molding temperatures will range up to, and over, 300°F. The cure time or molding cycle is dependent on the resin system, molding temperature, and the thickness, size, and shape of the item being produced and ranges from 30 seconds to 5 or more minutes.

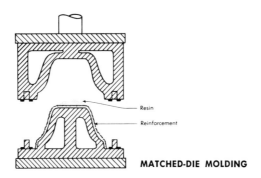

Resin

Reinforcement

MATCHED-DIE MOLDING

CONTINUOUS LAMINATING

The majority of flat and corrugated sheet for architectural applications is made by this method. Fabric or mat, or a combination of reinforcements, is fed from rolls, passed through a resin bath, and brought together to form the composite layer between two cellophane covering sheets. The lay-up thickness and resin content is controlled by exerting pressure on the squeeze rolls through which the laminate passes. The laminate is drawn through a heating zone by a suitable pulling mechanism. Room temperature cures are not used.

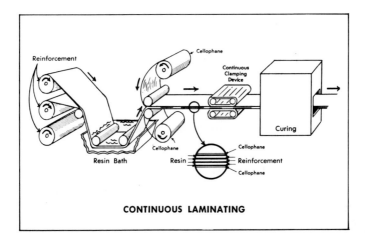

CONTINUOUS LAMINATING

5

Resins and Resin Systems

Resins used with fibrous reinforcements to produce reinforced plastics serve a number of purposes. Although the reinforcement provides the structural strength, the resin maintains the desired orientation of the reinforcement, provides for stress transferral from one fiber to another, and provides an enclosing armor around each fiber to keep it separated and to prevent abrasion from adjoining fibers. In addition the resin imparts corrosion resistance to the system and contributes to its thermal and electrical properties. Varying the ingredients in the resin system and careful selection of the reinforcement make it possible to provide for a wide range of end uses.

THERMOSETTING AND THERMOPLASTIC

Two main classes of resins are used in reinforced plastic products: "thermosetting" and "thermoplastic." Thermosetting resin systems become hard when cured or when heated, and further heating will not soften them—the hardening is irreversible. Thermoplastic resins become soft when heated and hard when cooled—the cycle is repeatable.

Thermosetting resins undergo a chemical change or reaction called polymerization, the linking of "monomers" to form "polymers." Time and temperature in the presence of catalysts or curing agents accomplish this reaction.

As mentioned earlier, the two thermosetting resins primarily used for reinforced plastics applications are polyesters and epoxies.

POLYESTERS

Polyesters, the workhorse of the industry, account for about three-fourths of the thermosetting resins used in industry and about 90 percent or more of the resins used in the reinforced plastics industry. The word *polyester* is derived from two chemical processing terms, *polymerization* and *esterification*. Technically speaking, polyesters are derived from a production process which involves the condensation esterification of dihydric alcohols and dicarboxylic acids.

This reaction covers a wide range of starting materials and can be extended and varied by the substitution or addition of other appropriate starting raw materials. Alteration of the chemical structure by processing techniques and raw material selection is a method of achieving the desired properties in the formulated resins.

Because of the versatility of polyesters and their capability to be modified or tailored (during their chemical construction), they have almost an unlimited range of uses in nearly every type of industry.

Polyesters offer the advantage of a balance of good mechanical properties, chemical and electrical properties, dimensional stability, low cost, and ease of handling.

Initially one can classify polyester resins into two major types: general purpose resins and special purpose resins.

The general purpose polyester is a low-cost polyester, with good electrical and mechanical properties and reasonably good corrosion resistance. It must be remembered that chemical resistance is a matter of degree and that even the least resistant polyesters are resistant to a wide range of chemicals and selected solvents.

General purpose resins are usually supplied in the medium- or low-viscosity range and in most cases require only the addition of a recommended catalyst to have a complete workable resin system. One of the largest uses of general purpose polyesters is for "open mold" work. These resins also find wide use in reinforced plastic luggage, trays, boxes, furniture, automotive components, and boat hulls—all applications which do not require outstanding corrosion resistance.

Special purpose polyesters are those resins which have been specifically formulated to meet some particular end-use requirement with respect to performance.

A. deDani in his book, *Glass Fibre Reinforced Plastics* (Interscience Publishers, Inc.) states that polyesters may be grouped under the following headings:

1. General purpose resins.
2. Plasticizing resins (flexible vs. rigid; preparation of gel-coats).
3. Thixotropic resins.
4. Resins for translucent sheets.
5. Chemically resistant resins.
6. Self-extinguishing resins.

7. Electrical grade resins.
8. Heat-resistant resins.
9. Resins for export via the tropics.

Oleesky and Mohr, in a more recent publication, *Handbook of Reinforced Plastics of the SPI* (Reinhold Publishing) suggest the use of five major classes to characterize polyester resins:

1. General purpose resins.
2. Light-stable and weather-resistant resins.
3. Chemical-resistant polyester resins.
4. Resins with high heat deflection temperatures.
5. Flame-resistant resins.

The *Modern Plastics Encyclopedia 1967* (McGraw-Hill) characterizes polyester resins with respect to uses as follows:

1. General purpose.
2. Light stabilized.
3. Surfacing.
4. Lay-up.
5. Chemical-resistant.
6. Heat-resistant.
7. Resilient.
8. Flame-resistant.
9. Flexible.
10. Special.

In the case of classification 10, "Special" indicates a resin "tailor-made for specific end uses."

More detailed discussion on characterization of polyester resins can be found in the three publications mentioned above.

Considering the above and the fact that there does not appear to be a system of uniform classification or characterization, it is important to be aware that:

1. The polyester resin industry is ready, willing, and able to supply polyester resins with widely divergent properties to satisfy the particular requirements to be encountered in the end-use application.
2. Polyester resins can be formulated or tailored to suit the requirements of the particular process in which they are going to be used and to emphasize a desired end property.

3. The formulation of special purpose polyesters involves the appropriate utilization of all the ingredients of the system, from the raw materials used to produce the basic resin, the reactive monomer for dilution and viscosity control, to the catalysts and accelerators involved in the cure.
4. Boundary lines of performance are not firm lines of demarcation. For example, a corrosion-resistant or chemical-resistant resin may also exhibit good temperature-resistant properties.
5. Much of the versatility of polyester resins results from the wide selection of raw materials, the basic processing methods, and the experience of the manufacturer in supplying the resin system that will meet the required product characteristics.

REACTION MECHANISM—POLYESTERS

This discussion is concerned with resin systems commonly referred to as "unsaturated" (reactive) polyesters. These systems are mixtures of true esters (long chain polymers chemically constructed in the processing reaction) dissolved in a polymerizable monomer (such as styrene) which provides cross-linking units to unite the chain three dimensionally. The two components (ester and monomer) coreact or copolymerize upon the introduction of a free radical donor catalyst (peroxide) to form a rigid, infusible thermoset material.

The term *unsaturated* (reactive) means that the unbroken double bonds (chemical linkages) are carried over from the original (acid) ingredients into the formulated polyester resin to provide points of reactivity (sites for cross-linking). In other words, the double bonds (sites of unsaturation) are opened up by the free radicals donated by the catalyst and accordingly unite with similar reactive chemical groups or units of the monomer.

The curing reaction is termed "addition" polymerization because no by-products result. In contrast, in phenolic resin systems, the curing reaction is termed "condensation" polymerization and a by-product of water is produced.

One should be aware of the "saturated" (unreactive) polyesters that are in common use in the plastics industry. These thermoplastic or linear polyesters are also referred to as "nonconvertible," since they do not convert into an infusible solid on heating. Accordingly, they should not be confused with the "unsaturated" polyester which is of prime importance to the discussion in this text.

To help in understanding the basic reaction of materials used to make polyesters and to further understanding of the steps of the reactions previously discussed, the following flow sheets and reactions are presented. One should realize that the reactions have been simplified and are not presented to illustrate the latest science and technology in the manufacture of polyester resins.

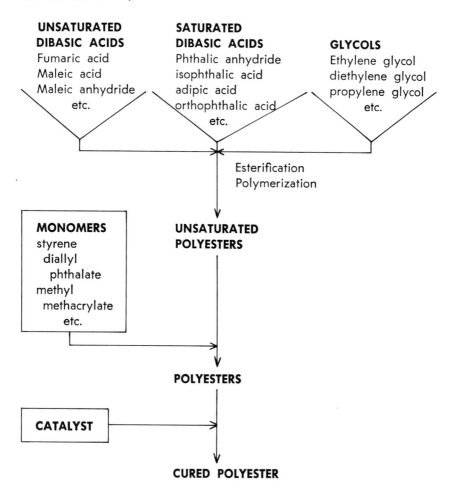

The flow sheet can be reduced to structural chemical formulas to further illustrate the reaction:

| Organic Alcohol | + | Unsaturated Dibasic Acid | (Esterification) $\longrightarrow$ | Ester + Water |

| Ethylene Glycol | + | Maleic Acid | (Esterification) $\longrightarrow$ | Ester + Water |

$$HOCH_2CH_2OH + HOOCCH{=}CHCOOH \rightarrow HOCH_2CH_2OOCCH{=}CHCOOH + HOH$$

62 pounds + 116 pounds $\longrightarrow$ 160 pounds + 18 pounds

The water is removed from the reaction by heat and condensation. Further addition of ethylene glycol would proceed as a polyesterification reaction:

$$HOCH_2 CH_2 OOCCH=CHCOOH + HOCH_2 CH_2 OH \longrightarrow$$

$$HOCH_2 CH_2 OOCCH=CHCOOCH_2 CH_2 OH + HOH$$

The reaction can proceed further with acid and glycol to give a long chain linear polyester which would be illustrated and further simplified:

$$HOC_2 H_4 OOCCH=CHCOOC_2 H_4 OOCCH=CHCOOC_2 H_4 \ldots . \text{etc.}$$

The above linear polyester is said to be unsaturated (reactive) at each group (CH=CH) since these groups have double bonds (=).

The reaction of the above resin with a monomer like styrene occurs at the reactive double bonds and may be depicted as follows:

$$HOC_2 H_4 OOCCH\text{-}CHCOOC_2 H_4 OOCCH\text{-}CHCOOC_2 H_4 \ldots . \text{etc.}$$

$$\begin{bmatrix} CH_2 \\ | \\ HC\text{-}C_6 H_5 \end{bmatrix}_n \quad \text{Styrene} \quad \begin{bmatrix} CH_2 \\ | \\ HC\text{-}C_6 H_5 \end{bmatrix}_n$$

$$HOC_2 H_4 OOCCH\text{-}CHCOOC_2 H_4 OOCCH\text{-}CHCOOC_2 H_4 \ldots . \text{etc.}$$

With the addition of the catalyst to the system, the cross linking unites the chain three dimensionally which results in a thermoset product (cured polyester resin).

Completion of curing of the above reaction after the addition of the catalyst is dependent on the formulation, the amount of the catalyst, and the time and temperature balance selected for the formulation. Pressure is not necessarily required for curing, but it can have an effect on factors such as surface smoothness, density (ratio of glass to resin), and other fabrication process considerations. One can demonstrate that this polymerization takes place by adding catalyst to the polyester resin (using a small quantity).

The fact that the styrene monomer enters into the reaction can be demonstrated by adding a small amount of catalyst to a small quantity of styrene. This catalyzed styrene will set up (harden) into a crumbly mass and will exhibit little in the way of strength properties. This experiment does not prove, however, that styrene is not a necessary and desirable component in a polyester resin system.

In working with polyester resin systems, the cure proceeds in two distinct stages. The first is the formulation of a soft gel. Immediately after this step occurs (gelation), the second phase of the cure takes place rapidly with considerable evolution of heat (exothermic reaction). In most cases complete cure is obtained without liberation of volatile materials.

The resin mix may contain materials other than the glass reinforcement. The ingredients, in addition to the monomer and catalyst, may consist of fillers, activators, inhibitors, pigments, lubricating agents, or waxes to provide tack-free cures.

MONOMERS

Polyesters as solid materials are blended with a reactive monomer to provide a workable resin material. Polyester resins as normally supplied by the manufacturer contain monomer, but some polyesters may be purchased without monomer. Additional quantities of monomer may be added to the resin at the user's plant (in amounts recommended by the manufacturer) to reduce the viscosity.

Styrene is the ideal monomer and the most commonly used monomer in polyester resin systems because it is abundantly available, low in cost, has good solvent characteristics, and reacts readily with the unsaturated polyester during the curing reaction.

Other monomers can be used in polyesters to provide for different end-use properties in the cured resin system. Methyl methacrylate is used with styrene in polyesters intended for exterior application to provide good weather resistance and good color stability. Methyl methacrylate can also be used to lower the refractive index of the resin system to approach the matching of the refractive index (1.548) of the glass reinforcement but by itself results in very poor cures in polyesters. This technique of matching refractive indexes is used to improve the transparency characteristics of the finished laminate.

The addition of excessive amounts of styrene monomer lowers the weathering performance of the product. Optimum weathering properties are demonstrated when equal quantities of styrene and methyl methacrylate are used in contrast to straight methyl methacrylate.

Vinyl toluene and diallyl phthalate are also commonly used as monomers although the relative high cost of the latter tends to limit its use.

Chlorostyrene is a more recent addition to the monomer field. This material, relatively high in cost and limited in availability, can be used to enhance selected properties of the polyester resin system such as high reactivity, improved surface characteristics, and craze resistance.

Additional monomers that have been used, but not in quantities comparable to styrene and vinyl toluene, are alpha methylstyrene, divinyl benzene, and triallyl cyanurate.

More technical information can be obtained in an article by Arthur L. Smith, "Monomers," in *Industrial and Engineering Chemistry,* 58, no. 4 (April 1966): 50–52.

In summary the main functions of a monomer are:

1. To act as a solvent carrier for the unsaturated polyester.
2. To modify the viscosity.
3. To enhance selected properties with respect to the end-use application.
4. To provide a rapid means of reacting with the unsaturated linkages in the polyester to yield a completed reaction and cross-linked copolymer.

The selection of monomers should be left to the polyester resin manufacturer. The laminator should generally limit the use of monomer addition to viscosity control and would be well advised to follow closely the recommendations of the manufacturer.

CATALYSTS

Organic peroxides are widely used to catalyze polyester resin systems and to initiate the copolymerization reaction. More technically, the peroxides decompose to release free radicals. These free radicals are attracted to the points of unsaturation (reactive sites) in the monomer used to dilute the polyester, attack these active sites, and initiate the polymerization reaction.

Three of the most common catalysts in use are methyl ethyl ketone peroxide, benzoyl peroxide, and cumene hydroperoxide. Organic peroxides are popular because of their convenience and cost. In addition, the use of organic peroxides facilitates control of the reaction (time of gelation and cure).

Polymerization or cure of polyester resin systems can also be achieved by exposure to radiation, ultraviolet light, and heat. Some of the factors regulating catalyst choice are:

1. Desired temperature, allowable time of laminating operation.
2. Type of monomer or mixture of monomer employed.
3. Desired pot life.
4. Desired gel and cure times.
5. Influence of sunlight and weathering on product performance.
6. Other specialized requirements of the finished end product influenced by the catalyst choice.

Over thirty catalysts are listed as being commercially available from over a dozen recognized suppliers.

Since catalysts may vary in strength because of type or age, it is a good idea to conduct a test with a small quantity of the resin system before proceeding with larger quantities.

The handling and storage of peroxides is discussed in chapter 24.

INHIBITORS

Inhibitors are added to or contained in polyester resins to enhance their storage stability. Hydroquinone is a commonly used inhibitor. The actual mechanism of inhibition is not completely clear.

Due to their reaction with free radicals from catalysts, or their reaction with active polymer growth centers, inhibitors prevent polymerization until they are consumed by some of the free radicals, after which polymerization proceeds normally.

Some inhibitors are effective with polyesters under storage as well as polymerizing conditions. The more popular types of inhibitors are those whose effectiveness is canceled by the introduction of heat or by other conditions which induce polymerization.

In summary, chemical inhibition is used in one or more of the following phases of polyester manufacture or usage:

1. To prevent premature polymerization during the manufacturing (esterification) process.
2. To provide stability in storage after manufacture.
3. To prevent premature gelation in intermediate or end-use processes such as mixing-milling or prolonged elevated temperature handling.
4. To stabilize monomers.

The laminator should consult with the resin supplier if the use or need for an inhibitor is indicated in his particular application.

Tertiary butyl catechol and hydroquinone, two of the more common inhibitors, are used with polyester to inhibit or slow up the gelling action. When the user adds inhibitor to the system, it is generally necessary to modify the amount of catalyst used in the resin system to obtain the desired results. Here again, a preliminary test of the resin system is suggested.

ACCELERATORS (Promoters)

Accelerators are usually added to polyester resin formulations to initiate or speed the gelling at room temperatures. Many resin suppliers offer fully pro-moted resins that require only the addition of the catalyst by the user to complete the cure. The recommendations of the manufacturer with respect to catalyst usage should be followed carefully. When used in combination with catalysts, a rapid curing reaction can be started without applying heat. Commonly used accelerators are cobalt naphthenate, diethyl aniline, and dimethyl aniline. The presence of cobalt naphthenate in a polyester resin system is generally indicated by a purple cast to the liquid resin. Accelerator and catalyst combinations must be carefully selected, one specific to the other, since only pairs or selected combinations will perform satisfactorily. Catalysts and accelerators should never be mixed together and then added to the resin system. See discussion in chapter 15.

It is desirable to have knowledge of the performance of the accelerator since its use may measurably shorten the "pot life" of the resin and not provide sufficient time for the resin mixing and the wet-out of the reinforcement prior to the beginning of the hardening of the resin system.

Fillers and Pigments

Fillers and pigments are added to molding resins to: reduce shrinkage, minimize crazing, lower material costs, impart color or opacity, and improve surface finish.

Excessive use of fillers will reduce the fluidity of the resin to a point where it will be difficult to handle. Use of fillers can make inspection for internal flaws difficult. Proper use—to accomplish definite objectives—can serve to an advantage.

Calcium carbonate, diatomaceous earths, and clays are commonly used fillers. Noncompact materials such as colloidal silicas, bentonite, mica platelets, or short-length fibers of asbestos or glass can be used with polyester resins to produce a flow characteristics known as thixotropy. The particles in these materials, because of their high surface-to-mass ratio, web together and coagulate the resin system to produce a stiff or jellylike consistency when at rest. When disturbed by agitation or otherwise subjected to stress, the polyester resin regains fluidity temporarily to facilitate application to a surface. The practical application of thixotropic agents is to permit the use of resin systems on vertical surfaces without excessive drain or runoff. Many resin manufacturers offer polyesters with the desired thixotropic properties. The addition of 2 percent to 3 percent by weight of colloidal silica to a polyester resin will make it resemble grease in consistency. Cab-o-Sil (a trademark of Cabot Corporation) and Santocel (trademark of Monsanto) are effective thixotropic agents.

Mold Release

Easy release of the part from the mold after completion of the curing cycle is normally accomplished by lubricating the mold with waxes or silicones. At other times, such substances as zinc stearate or paraffin wax are added to the resin mix in order to facilitate the removal of parts. The latter are called internal lubricants.

Other Additives

Other materials may be added to the resin mix either by the resin supplier or the molder to impart special performance qualities. Typical of these substances are ultraviolet absorbers which are added to resin mixes subject to exposure to ultraviolet rays in natural sunlight or fluorescent light.

Flameproofing substances, such as antimony trioxide or chlorinated waxes, are added to achieve a fire-retardant property. In the case of fire-retardant additives, their effect on outdoor weathering performance must be carefully considered since some of the halogen-type additives will impart discoloration to a laminate after it has been exposed for a reasonable period of time.

EPOXIES

Epoxy resins are so named because they are made from epoxides or compounds containing the epoxy or oxirane structure:

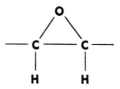

Epoxy resins are viscous liquids or brittle solids which become commercially useful structural materials when they are reacted or hardened (cured) with other materials. Epoxies like polyesters are thermosetting types of resins and are reacted with curing agents or catalytically homopolymerized to form a cross-linked polymeric structure. The preponderance of available literature deals with the cure of epoxies rather than the preparation of epoxies.

The commercial development of epoxies began in Germany in the 1930's. German Patent 676117 (1939) describes liquid polyepoxides (compounds containing two or more epoxy structures) which can be hardened by a variety of methods. Here in the United States the earliest disclosure was U. S. Patent 2,324,483 (1943) and it relates to the curing of epoxy resins with dibasic acids.

From a manufacturing standpoint, the most common process for producing epoxies is the reaction of epichlorohydrin with bisphenol-A in aqueous caustic soda.

Since their introduction, epoxy resins have been used increasingly; over 150 million pounds were sold in 1966. The cured epoxy resin systems are characterized by low shrinkage during cure, low water absorption, and excellent corrosion resistance. Another outstanding characteristic is their excellent adhesion to other surfaces.

It is claimed that epoxy systems exhibit better corrosion resistance than polyesters to solvents and alkalis, but poorer resistance to acids. Epoxies exhibit good mechanical strength (high strength-to-weight ratio), good electrical properties, good dimensional stability, and good thermal insulation properties.

Their excellent thermal properties make them ideally suited for elevated temperature applications (normally up to 300°F and in specially formulated systems they have performed up to 600°F).

Of the many epoxy resin types, those of more important commercial significance are the condensation product of epichlorohydrin and bisphenol–A, utilizing a cure system of amines, acid anhydrides, or latent type curing agents. Out of several hundred organic and inorganic chemicals that have been investigated as curing agents or catalysts for cross-linking (curing) epoxy resins, only about 50 are used in commercially significant quantities.

Like the polyester resin system, the successful end-use application of an epoxy resin system is dependent on the selection of the proper epoxy-resin curing system, the curing conditions, and the correct processing methods. This also means that the appropriate reinforcement and/or filler has been selected.

The superior properties of the epoxy resins, as contrasted with polyester resins, tend to compensate for their higher cost. In the past, the use of epoxies for general types of reinforced plastics was limited to critical applications where the materials' superior properties were required. But with the development of filament-wound structures and the advantages of using epoxy resins, reinforced plastics are currently finding many new end-use applications.

Processibility and properties of a reinforced epoxy material as previously mentioned are highly dependent on the type of hardener used to cure it. Aliphatic amines provide optimum handling characteristics. Aromatic amines provide good short-time heat resistance. To develop more flexibility, producers use vegetable oil polyamides as the curing agent or hardener. For applications requiring greater temperature resistance, specific anhydrides or dianhydrides are indicated.

Catalysts or accelerators as well as pigments are compatible additives to epoxy resin systems. In general these additional materials serve the same purpose for epoxies as they do for polyesters. Reactive diluents are used with epoxies to afford viscosity control whereas monomers are used with polyesters to vary the properties of the curred system and control viscosity.

Epoxy systems can be formulated with selected curing agents to cure at room temperature; but more commonly heat is used to adequately cure the epoxy system, and postcuring is used particularly if the user is desirous of obtaining the ultimate in mechanical properties. Similarly, many polyester systems can be cured adequately at room temperature, but heat can be used to speed the production process and postcure can be used to improve mechanical properties.

It is common practice in the reinforced plastics industry to use epoxy-type adhesives to join (bond) polyester glass-reinforced laminates in the fabrication of large structures or components. The use of epoxy materials for adhesive applications accounts for sizable quantities of the total consumption.

SELECTION GUIDE—POLYESTERS

See the following page for a guide to the various characteristics and typical uses of polyesters.

SELF-EXTINGUISHING POLYESTERS
AND EPOXIES

Self-extinguishing polyesters are made from a number of compounds which are either additives to, or integral parts of, the chemical structure of the resin system. Included among these materials are chlorinated paraffins, tetrachlorophthalic anhydride, tetrabromophthalic anhydride, trichlorethyl phosphate, "HET" (trade name of Hooker Electrochemical Co. for hexachloroendomethylene tetrahydrophthalic anhydride), or chlorendic anhydride, chloro maleic acid, organic phosphonates, and chlorostyrene. Antimony trioxide added to polyester in small amounts, 1 to 5 percent by weight, is effective in increasing fire resistance, particularly when used in combination with a halogen-containing polyester system.

Some self-extinguishing polyesters have a high viscosity which makes them somewhat difficult to handle. Compounds added to increase fire resistance, which are not chemically bound in the resin, may reduce the physical properties and leech out upon exposures. The effect of these modifying agents on the strength and other properties of the laminate should be carefully investigated and calibrated prior to use in a specific application. Flame-retardant polyester laminates tend to have poor weathering characteristics and turn yellow or deteriorate when exposed to sunlight. The use of light screening agents (U-V stabilizers) has helped to minimize these poor weathering characteristics.

One method suggested for increasing the fire resistance of a laminate is to seal its surface with a self-extinguishing resin. The interior of the laminate is made with ordinary polyesters but its outermost plies are laid up with a self-extinguishing type.

Monomers should be added to self-extinguishing polyesters only when recommended by the resin supplier and then never in more than the recommended amount.

Epoxy resins formulated with diglycidyl either of tetrachlorobisphenol-A, and cured with chlorendic anhydride as a hardener, qualify in the fire-retardant category. The addition of 5 percent antimony trioxide is also practiced to further enhance the fire-retardant properties.

FIRE RETARDANCY

There are many applications in industry where reinforced plastics are acceptable for use and can serve to good advantage over other construction materials. Lack

	Polyester	Characteristics	Typical Uses
Named by Characteristic of Cured Resin	General purpose	Rigid moldings, craze-resistant	Trays, boats, tanks, boxes, seating
	Flexible resins and semirigid resins	Tough, good impact resistance, high flexural strength, low flexural modulus	Vibration damping: machine covers and guards, safety helmets, electronic part encapsulation, patching compounds, auto bodies, boats
	Light stable and weather-resistant	Resistant to weather and ultraviolet degradation	Structural panels, skylighting, glazing
	Chemical-resistant	Highest chemical resistance of polyester group, excellent acid resistance, fair in alkalies	Corrosion-resistant applications such as pipe, tanks, ducts, fume stacks
	Flame-resistant	Self-extinguishing, rigid	Building panels (interior), electrical components, fuel tanks
	High-heat distortion	Service up to 425°F, rigid	Aircraft parts
	Electrical	Fast cure, good hot strength, good electrical properties	Where good electrical properties are required
Named by Processing Characteristics	Hot strength	Fast rate of cure, "hot" moldings easily removed from die	Containers, trays, housings
	Low exotherm	Void-free thick laminates, low heat generated during cure	Encapsulating electronic components, electrical premix parts—switchgear.
	Extended pot life	Void-free and uniform, long flow time in mold before gel	Large complex moldings
	Air dry	Cures tack free at room temperature	Pools, boats, tanks
	Thixotropic	Resists flow or drainage when applied to vertical surfaces	Boats, pools, tank linings

of complete understanding, together with lack of standardized tests, and lack of their acceptance, points up the need for more meaningful information on the fire retardancy of reinforced plastic materials. Flame resistance and flammability are terms which are often loosely used. The terms are meaningless unless they can be expressed in units that are based on accepted flammability tests.

The following is presented to provide a better understanding of the tests that are currently used to evaluate the flammability of plastic materials.

Surface Burning Characteristics of Building Materials
Underwriters' Tunnel Test (ASTM E 84-70,
Pt. 14, 1973 ed.)

ASTM Standards state, "this method for surface burning characteristics of building materials is applicable to any type of building material that, by its own structural quality or the manner in which it is applied, is capable of supporting itself in position or may be supported in the test furnace to a thickness comparable to its recommended use. The purpose of the test is to determine the comparative burning characteristics of the material under test by evaluating the flame spread over its surface when exposed to a test fire and to establish a basis on which surface burning characteristics of different materials may be compared, without specific considerations of all end-use parameters that might affect the surface burning characteristics. Fuel contributed and smoke density as well as the flame-spread rate are recorded in this test. However, there is not necessarily a relationship among these three measurements. It is the intent of this method to register performance during the period of exposure, and not to determine suitability for use after the test exposure."

The test chamber is a horizontal duct 17½ inches wide, 12 inches high, and 25 feet long. Red oak is the calibration standard and is arbitrarily assigned a value of 100. A value of 0 is assigned to asbestos. Other materials are reported proportionally.

The following table lists some typical values that have been obtained in the tunnel test:

Material	Flame-Spread Ratings
Asbestos Cement Board	0
Non–Com Fire Protected Wood[1]	15 or less
Fire Retardant Polyester Panels	20–75
Red Oak Flooring (Select grade)	100
White Pine	130
Plywood	100–180
General Purpose Polyester Laminates	150–400
Veneered Wood	515

1. Non–Com is a registered trademark of Koppers Company, Inc.

Detailed information on the use of different materials in applications where flame-spread ratings control the use of the material can be obtained by consulting the various codes and agencies. Codes applicable to different geographical locations control the quantities of materials that may be used in buildings as related to their respective flame-spread ratings, or the approved test for rating the flammability of the material.

It has been suggested that the following classifications can be assigned to the various flame-spread ratings:

Flame-Spread	Classification
0–25	Noncombustible
25–50	Fire Retardant
50–75	Slow Burning
75–200	Combustible
Over 200	Highly Combustible

There is no question that the Underwriters' Tunnel Test is a good measure of the surface burning characteristics (flame spread) of a specific material. Further it establishes a basis on which surface burning characteristics of different materials may be compared. The test is expensive to set up and to conduct. It has limited usefulness for screening determinations of flame-retardant materials.

Surface Flammability of Materials Using a Radiant Heat Energy Source (ASTM E 162–67, Pt. 14, 1973 ed.)

ASTM Standards state that "this method, to be used for research and development purposes, covers the measurement of surface flammability of materials. It is not intended for use as a basis of ratings for building code purposes."

Feuer and Torres of Atlas Chemical Industries reported that, in their analysis, the Radiant Panel Test appeared to be the best method upon which to concentrate their efforts.[2] The test panels and test conditions tend to duplicate actual service conditions and offer reasonably good reproducibility.

Flammability of Self-Supporting Plastics[3] (ASTM D 635–74, Pt. 35, 1974 ed.)

ASTM Standards state, "This method covers a small-scale laboratory screening procedure for comparing the relative flammability of self-supporting

2. *Chemical Engineering* (April 2, 1962): 139–142.
3. ASTM makes the following admonishment: *"This standard should be used solely to measure and describe the properties of materials, products, or systems in response to heat and flame under controlled laboratory conditions and should not be considered or used for the description, appraisal, or regulation of the fire hazard of materials, products, or systems under actual fire conditions."*

plastics in the form of bars, molded or cut from sheets, plates, or panels, tested in the horizontal position. This method should be used to establish relative burning characteristics of plastic materials and should not be used as a fire hazard test method."

Incandescence Resistance of Rigid Plastics[4]
(ASTM D 757-74, Pt. 35, 1974 ed.)

ASTM Standards state, "This method of test provides for laboratory comparisons of the resistance of rigid plastics to an incandescent surface 950 $\pm$ 10°C (1742 $\pm$ 18°F). It may supplement tests using a flame source of ignition, such as Method D 635."

Flammability of Plastics Using the Oxygen Index Method[5]
(ASTM D 2863-74, Pt. 35, 1974 ed.)

ASTM Standards state, "This method describes a procedure for determining the relative flammability of plastics by measuring the minimum concentration of oxygen in a flowing mixture of oxygen and nitrogen that will just support flaming combustion. This method has been found applicable for testing various forms of plastics materials including film and cellular plastic."

Oxygen index is defined as "the minimum concentration of oxygen, expressed as volume percent, in a mixture of oxygen and nitrogen that will just support flaming combustion of a material initially at room temperature under the conditions of this method."

The T-341: A Method to Measure Self-
Extinguishing Characteristics of Reinforced
Plastic Duct Under Severe Burning Conditions

McMahon and Slama of The Ceilcote Company, Inc., have published a test method for measuring the self-extinguishing characteristics of chemical-resistant reinforced plastic duct in accelerated burning conditions. The T-341 test simulates and greatly accelerates actual conditions of a fire in duct work being fed by a forced draft.

"This test (T-341) shall be used to determine whether a material is self-extinguishing under the simulated operating conditions of a forced draft and high intensity heat source. The test specimen is rated self-extinguishing if it snuffs out in 45 seconds. The test specimen shall be a 10 foot section of 8 inch diameter duct, 3/16 inch wall thickness."

The authors conclude: "1. The T-341 fire test better simulates fire-in-ductwork conditions than tests now available. 2. It therefore provides more reliable data."

4. Ibid.
5. Ibid.

The work of McMahon and Slama offers a meaningful method for evaluating flammability characteristics of reinforced plastic ductwork under severe burning conditions. Their efforts should be recognized as a valuable contribution in an area where more information is needed with respect to flammability of reinforced plastics.

Flame Resistance (Military Specification) (LP406 b No. 2023.2)

This test is designed to measure flame resistance of materials used in electrical equipment where arcing of the electrical current may cause a material to ignite. It is not practical to correlate the results of this test with other test results involving reinforced plastics.

Factory Mutual Calorimeter Test

This test causes complete combustion of the material and accordingly does not demonstrate the ability of a plastic material to extinguish itself when the flame is removed. In view of this, it cannot be used as a measure of flame spread.

Factory Mutual "Whitehouse" Test

This is another test that has been developed to measure the flame spread of materials. It is comparatively expensive to conduct.

HLT-15 Intermittent Flame Test

This is a relatively simple test requiring a minimum of apparatus. It is considered to be a more severe test than some of the others described since it is conducted with the specimen suspended in a vertical position. Heat from the burning will be carried upward by convection along the length of the specimen.

In rating a composition, five specimens are burned and each specimen is ignited five times using increasingly longer ignition periods. If the flame is extinguished within the period that the burner flame is withdrawn, the specimen has passed the ignition test. The test method calls out the ignition time in seconds and the specified time for extinguishment is twice the ignition time. This test should be considered in the light of a laboratory method to evaluate the burning rates of somewhat similar compositions.

The more important flammability tests are those which can classify the material being tested with respect to its fire hazard. All of the common organic plastics, including the fire-retardant types, will combust when exposed to a direct flame. At the present time no universally accepted test has been developed to standardize plastic materials with respect to fire retardancy. Additional work is needed in this area.

Three requirements are fairly well agreed upon for materials that may become exposed to industrial fires:

1. The material should not contribute to the transfer of a fire from one location to another.
2. The products of combustion should not interfere with the extinguishing of an existing fire.
3. The quantity of smoke developed should be minimal.

6

Reinforcing Materials

Drawn glass fiber is the major reinforcing material used in reinforced plastics in corrosion-resistant applications. Asbestos fibers, as well as organic fibers, are also used in corrosion-resistant laminates or components (such as pipe and fittings). The reinforcements in use can be divided into five classes: (1) metal reinforcements, (2) inorganic fibers, (3) wood and vegetable fibers, (4) paper, and (5) natural or synthetic organic fibers.

GLASS

Most glass-reinforcing materials are made from two grades of glass which differ in chemical analysis—"E" and "C" glass. "E" grade is best applied for structural uses where high heat resistance and superior electrical properties are desired, and "C" grade for uses where superior resistance to chemical corrosion is required.

There are two basic forms of fiber glass reinforcements—continuous filaments, generally made from "E" glass, and staple fibers, generally made from "C" glass. Both forms of fiber glass begin with the same manufacturing process, but differ in the manner in which they are drawn from the glass furnace.

A continuous filament is an individual fiber of a continuous filament strand. A strand, composed of many continuous, fine filaments—from 51 to 408—is drawn through platinum bushings from the furnace at speeds of 100 to 200 feet per second. The investment in platinum bushings in a large glass plant may exceed one million dollars.

A staple fiber is an individual fiber 8 to 15 inches long, formed by jets of air, which pull the glass filaments from the furnace. They are gathered and held by suction on a revolving perforated drum, later to be rewound as an untwisted rope. During the pulling or drawing operations, sizing material is placed on the filaments. Surface treatments (sizings and binders) are discussed later.

Continuous filament and staple fibers can be fabricated into yarns and cords through conventional twisting, plying, and cabling operations.

Fiber glass reinforcements are supplied in several basic forms, which allow for flexibility in cost, strength, and choice of process. Many variations of the basic forms have been developed over the years to meet the ever-increasing performance and economic needs.

TRANSPORTABLE FORMS

The basic transportable forms of fiberglass for reinforcement of resin systems are:

1. **Continuous strand.** Two basic forms of continuous strand are available—yarn and roving. Roving is available in continuous filament and spun-strand construction.
2. **Mats.** Two basic types of mat are available—reinforcing mat and surfacing mat.
3. **Chopped strands.** Produced by chopping continuous strand roving into ¼-, ½-, 1-, 1½-, 2-, or 3-inch lengths. When produced from spun roving, lengths are generally shorter and the maximum length is the controlling specification. Milled fibers are hammermilled into 1/32 or 1/8 inch lengths.
4. **Fabrics and tapes.** These are made from continuous filament yarns, rovings, and staple yarns. Additional processing or weaving is required to make the reinforcement available in these forms.

Surface Treatments

Glass fibers require specialized treatments during the production process and subsequent use. Each treatment, generally organic, provides some specific beneficial quality such as providing lubrication, protecting the individual filaments, holding the strands together, or providing compatibility with the resin system used in the manufacture of the glass-reinforced end-product.

1. **Sizings.** These surface treatments generally can be classified as two basic types, "textile sizings" and "reinforcement sizings."

Textile sizings are ordinarily used for treatment of fibers to be used later for manufacture of fabrics.

Reinforcement sizings are generally planned for chemical treatments that will be compatible with further treatments (finishes) or that will improve the bonding properties of the glass fiber with the resin system.

The sizing material is applied to continuous filament fiber glass during the drawing operation at a point about 2 feet below the bushing. The application point is important to make sure that all 204 filaments are wetted by the size just prior to combining into a strand.

Sizing materials are of two formulations:

a. Textile Size—A complex starch-oil emulsion applied to fibers to be used for textile type applications (wrapping, twisting, plying, or weaving), and
b. Reinforcement Size—A chemical complex containing an organic resinous film former, a wetting agent, and a surface active agent applied to fibers intended for use in the reinforcement of resin systems.

The textile sizings can be removed (burned off) by subsequent heat testing or heat cleaning. This permits the addition of a finish in a subsequent operation to make the fiber compatible with the resin system to be used with the reinforcing material.

The surface active agents used in the chemical complex (reinforcement) size are generally of two types: (1) the chrome complexes; (2) the silanes which have been developed for use with glass fibers. In everyday parlance, the materials are commonly called chrome or silane sizes or finishes.

Textile sizes are generally used as lubricants during ensuing weaving operations and are chemically noncompatible with respect to the use of the glass with a resin system. In contrast, reinforcement sizes, chrome or silane, are used to improve the chemical bond between the molding resin and the glass filaments in the reinforcement.

2. **Finishes.** Finishes consisting of chrome or silane chemical complexes are applied to fabrics only after weaving and heat cleaning. Commercial heat cleaning or "carmelizing" removes the starch-oil emulsion type size by combustion processing, and is more efficient than chemical washing, which has also been used to remove sizings. "Carmelizing" describes the process of converting starch to carbon form under controlled thermal conditions. Where the fiber glass material (fabric or cloth) is intended for use in laminates, it is imperative that the fabric be heat-treated or chemically washed to remove the sizing, and that a chemically compatible finish be added. It is not uncommon to combine the heat-cleaning and finishing operations into one operational step.

For an in-depth study of fiber glass finishes, many recognized comprehensive texts are available.

The user or purchaser should contact his supplier for specific recommendations on finishes for his intended end-use application.

Leading producers of glass fibers use the designations listed below to describe finishes applied to reinforcements that are suitable for use with polyesters or epoxies. There are other finishes available for end-use applications which do not involve reinforced plastics. Some suppliers may, of course, use different designations for finishes which, in general, are chemically similar.

Finish 112 (Heat Cleaning) This is basically a treatment for use prior to the application of a finish in which almost all of the organic sizing is burned off the reinforcement. The residue is about 0.1 percent and the reinforcement has its original natural white appearance.

Finish 114 or **Volan**[1] After heat cleaning (Finish 112) the reinforcement is completely saturated with a chrome complex (Volan). This finish is recommended for use with polyester resins, but laminates made using this finish may be deficient in wet strength.

Volan[2] **A** This modification of Finish 114 provides improved wet strength. This finish is recommended by reinforcement manufacturers for polyester and epoxy laminates which require good wet- and dry-strength properties. Volan A Finish is obtained by saturating 112 Finish with methacrylate chromic chloride under controlled conditions so that the chrome content of the finished reinforcement is between 0.03 percent and 0.06 percent by weight.

Finish 136, 301, and **Garan**[3] Following heat cleaning, the fabric is completely saturated with a silane-type chemical and carefully dried. A vinyl-silane is left on the surface of the glass; it provides a chemical bond between the glass surface and the polyester resin. Fabrics so treated exhibit outstanding wet- and dry-strength properties, and are used with polyesters.

Finish A-172 A type of vinyl silane saturation treatment which exhibits good wet-out properties with polyesters.

Other finishes for use with epoxy resin systems are available for applications which demand high performance properties, particularly military applications.

After the chemical treatment is cured, the reinforcement is very thoroughly washed to remove certain undesirable soluble chlorides. This finish is recommended for use with polyesters, epoxies, and phenolic resins and improves the wet- and dry-strength properties of the laminate. This finish also provides improved drapability and wet out of the reinforcement during the laminating process.

Finish A-1100 (Silane Saturation) Following heat cleaning, the reinforcement is completely saturated with an amine-type vinyl silane and then dried. This finish is particularly well suited for use with epoxy resins or other temperature-resistant resin systems, such as phenolics, but is generally not recommended for use with polyesters.

3. **Binders.** Binders are either liquid emulsions, or solutions of resinous materials, or pulverized resin solids, and are used to chemically bond fibers in a fixed relationship. In producing fiber glass mat and surfacing veil for reinforcement applications, binders are directly applied to virgin blown fibers. In forming

1. Trademark of E.I. duPont de Nemours & Company, Inc.
2. Ibid.
3. Trademark of Johns-Manville Corporation.

mat products, binders are also applied to previously "sized" strands. Liquid and solid polyester resins and some liquid acrylic resins are used on the largest portion of fiber glass reinforcing mat products. When it is necessary to match binder resin properties with those of molding resins ultimately to be used, specific binders can be designated. The choices include soluble and insoluble binders.

In preform types of production operations, a polyester resin is commonly used to hold the chopped-strand preform fibers in fixed relationship until the preform is further processed in the matched-metal molding equipment. Here the binder serves to permit the handling of the preform (by making it transportable).

Types of Reinforcements—Glass

The general types or structures of glass-reinforcement products available commercially, in transportable form, are:

1. Continuous Strand
   a. Yarn
   b. Roving
      (1) Continuous filament-strand roving
      (2) Spun-strand roving

(PPG Industries)

(PPG Industries)

(Owens-Corning Fiberglas Corp.)

2. Mats
    a. Reinforcing mats
        (1) chopped-strand mat
        (2) continuous-strand mat
        (3) combination
    b. Surfacing mats

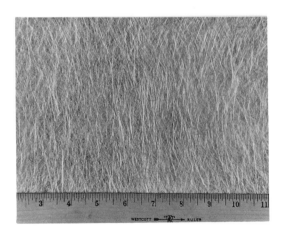

(Owens-Corning Fiberglas Corp.)

(PPG Industries)

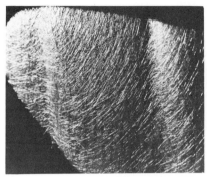

(PPG Industries)

(Owens-Corning Fiberglas Corp.)

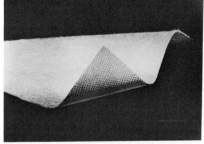

(Fiber Glass Industries, Inc.)

3. Chopped Strands

(Owens-Corning Fiberglas Corp.)

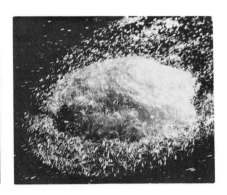

(PPG Industries)

(Johns-Manville Fiber Glass Reinforcements Division)

4. Fabrics and Tapes
    a. Continuous-filament fabrics
    b. Staple fiber yarn fabrics
    c. Woven roving fabrics
    d. Spun-strand roving fabrics

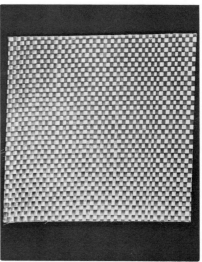

(Owens-Corning Fiberglas Corp.)

(Johns-Manville Fiber Glass Reinforcements Division)

(Owens-Corning Fiberglas Corp.)

(PPG Industries)

The material below further details the structure of commercially available fiber glass reinforcements.

Continuous Strand

A strand is composed of many continuous fine filaments (from 51 to over 400, depending on the specific requirements and the manufacturer). A continuous filament can approach infinite length (several miles).

a. **Yarn** (continuous filament) is made by twisting and/or plying a number of glass fibers to provide reinforcements of different strengths, diameters, and flexibility.
b. **Roving** is made by gathering a number of continuous filament strands and winding them on a cylindrical package.
   (1) **Continuous filament strand roving** is available in a wide range of weights, the most common being 240 yards per pound.
   (2) **Spun-strand roving** is a lower-cost roving consisting of one or more continuous strands looped back and forth upon themselves. They are held together by slight twisting and a cohesive sizing. This type of roving has less unidirectional strength than continuous strand roving, but it does possess greater bulking characteristics and is lower in cost.

Mats

Mats (nonwoven) are made from strands evenly distributed in a random pattern. Their basic application is to insure maximum uniformity in the finished laminate. Strands are held together by soluble or insoluble adhesive resinous binders or mechanically held in place by "needling." Where adhesive binders have been used, the user should make sure that the binder will be compatible with his resin system. Mats may be treated with various resins to provide optimum compatibility with the laminating resin system. Reinforcing mats are lower in cost than woven materials.

a. **Reinforcing mats**
   (1) Chopper-strand mat is constructed from multiple-length, chopped fine strands and is bonded in mat form with a resinous binder. Its principle application is for reinforcing polyester resins under contact molding pressure where fast wet-through, minimum resin absorption, and ease of forming are of prime importance. It is available in weights up to 3 oz. per square foot and in widths up to 72".
   (2) Continuous strand mat is made from multiple layers of continuous filament strands in a swirl pattern, which are held together with a resinous binder. This mat is recommended for matched metal die molding where relatively deep and complex contours require maximum draw characteristics. It is available in weights up to 3 oz. per square foot and in widths up to 48".
   (3) Fabmat, a trademark of Fiber Glass Industries, Inc., is a combination reinforcement consisting of one ply of fast-wetting Araton, a trademark of Owens-Corning Fiberglas Corp., woven roving, chemically bonded to chopped-strand mat. The two components

are bonded together with a highly soluble polyester powder binder to form a strong but drapable unit reinforcement that combines the bi-directional orientation of woven roving and the isotropic orientation of chopped-strand mat. Fabmat is available as a 16, 24, or 28 oz. per sq. yard woven roving bonded to 1.0 or 1.5 oz. per sq. foot chopped-strand mat. Fabmat saves time in lay-up as two layers are placed in the mold in a single operation.

b. **Surfacing mat** is commonly used with other glass reinforcements for laminate surface appearance and weathering. It is also used to provide reinforcement of a resin-rich surface where serious corrosion considerations are involved.

Standard surfacing veil or mat is a thin, highly porous mat made from monofilaments of type "C" glass, arranged in a veil-like pattern. By the very nature of its glass patterns, it has little reinforcing strength and accordingly is not designed to contribute to overall strength. The surfacing mats are available in thicknesses from $0.010''$ to $0.030''$. Surfacing mat serves to cover irregularities by drawing a slight excess of resin to the surface next to the mold. The smaller the diameter of the glass fiber used to produce the mat, the greater the improvement of surface irregularities. The mat or veil, in addition to promoting a smooth surface, compensates for shrinkage in a resin-rich surface.

c. **Overlay mat** is very similar to surfacing mat and is commonly used in matched-die molding to improve surface properties. It differs from surfacing mat because of its lower content of resinous binder. It is difficult to handle due to tearing of the mat structure.

Surfacing mats described above may be decorated by silk-screen printing or with a pigmented resin and used in combination with other glass reinforcing. The resin used to decorate the mat should be compatible with the resin that is used to produce the laminate. There is no limit to the variety of decorative effects that can be achieved.

Chopped Strands

Chopped strands are produced by cutting continuous-filament strands or spun strands. Chopped continuous-filament strands are used when uniform standard lengths are desired from $\frac{1}{4}''$ to over $3''$, while chopped strands made from roving are not uniform in length and lengths will range from $\frac{1}{4}''$ to $\frac{1}{2}''$ with the maximum length as the controlling specification. In preform fabrication techniques using chopped strands, the strands are continuously chopped on the job site immediately prior to use rather than purchased in chopped form. Milled fibers made by hammer milling continuous-filament strands are sometimes used as fillers or flow control agents. Chopped-spun strands are one of the lowest cost forms of glass reinforcement available.

Fabrics and Tapes

Fabrics, tapes, and woven rovings are made by interlacing (weaving) yarns of glass fibers of varying twist and ply construction on conventional weaving looms. Fabrics and tapes offer exacting control over thickness, weight, and strength. "Broad fabrics" refer to woven glass fabrics 18″ or wider, with the standard widths ranging from 38″ to 60″. Tapes are sometimes called "narrow fabrics" and refer to woven glass ranging in widths from ¼″ to 8″.

Industrial fabrics are normally woven from continuous-filament yarns, staple fiber yarns, and rovings. Tapes are woven from continuous-filament and staple fiber yarns. Woven rovings are manufactured from roving or spun roving, or a combination of both.

The essential variables in fiber glass fabrics are thickness (0.0010″ to 0.067″), weight per square yard (0.6 oz. to 27 oz.), construction, and widths (up to 60″). The basic characteristics of fabrics are controlled by the yarn weight and count plus the type of weave employed. Other factors being equal, the cost for fabric on a weight basis in a laminate becomes less when woven of fewer heavy yarns as opposed to many fine yarns, when woven with yarns having coarser filaments, and when the lowest number of plies of the heaviest satisfactory fabric is used.

Construction count is always quoted with the number of warps (lengthwise yarns) per inch first—followed by the number of picks (sometimes called wefts) per inch which run crosswise to the warps and constitute the fill. Generally speaking, low-pick constructions are somewhat less costly to weave than high-pick constructions. Typical construction counts in fabrics range from 100 ends warp to 80 picks fill down to 16 ends warp to 14 picks fill. Woven roving, of course, will exhibit even lower counts.

In addition to controlling weight and strength, the size of yarn used is important since it controls fabric thickness. The effect of yarn size on breaking strength and on weight is generally in direct proportion, with heavier yarns giving higher values.

Generally speaking, many layers of a thin fabric in a laminate will give higher physical test results than few layers or plies of a thick or heavy woven reinforcement. The ability to multiply strength values by using a multiplicity of laminations is one of the important advantages inherent in woven glass fabrics as a reinforcement for plastic materials.

It is important to remember that the strength of the fabric is directly proportional to the number of yarns per inch and the strength of the yarn selected. The lowest cost fabrics are those woven with the maximum number of yarns in the warp direction and the minimum number of fill yarns. To interlace additional fill yarns results in a more costly fabric than one made with a minimum of warp yarns.

Fabrics are carefully engineered products and the weaving of glass fabrics, along with the production of glass fibers, is in a continuing state of development. In light of this situation, which will probably exist for some time to come, it is especially advisable for new users of glass fabrics to consult with qualified suppliers in the selection of the most suitable and economic fabric to produce the desired results.

For specialized applications, braided fiber glass tubing or sleeving is also available. Yarn is also knitted on site in specialized operations to provide the optimum in reinforcement.

Three-dimensional weaving and knitting of reinforcing materials is common practice. With these techniques, contoured three-dimensional fiber glass reinforcement can be produced for fabrication into reinforced plastic parts or shapes. These specially tailored reinforcements should provide for improved strength properties in the third direction (shear) and overcome this deficiency in laminates that contain only two-directional reinforcement.

a. **Continuous-filament yarn fabric** (cloth) is available in open, medium, and close weaves. Weights range from 1 ounce to over 40 ounces per square yard. Thicknesses range from 0.0010″ to 0.040″. Continuous-filament plied yarns are normally used in weaving cloth fabrics, but single yarns are also used. This type of fabric is stronger and thinner than fabric woven from staple fiber yarns and provides high strength and flexibility with minimum thickness.

b. **Staple fiber yarn fabric** is known for its bulking and cushioning characteristics. It is available in medium weaves and varies in thickness from 0.015″ to 0.065″ with a weight range of 10 to 48 ounces per square yard.

c. **Woven-roving fabric** is manufactured from continuous-filament roving, spun-strand roving, or a combination of the two. It possesses a bulky coarse texture and as fabric offers the highest strength at lowest cost. It is available in plain or unidirectional weave. By its inherent structure, woven roving offers the fabricator a method of rapid build-up of thickness in a laminate preparation by a factor of about 3 compared to cloth.

When woven rovings are used in laminating operations, two types of wetting must be considered. These are wet-through (movement of resin through open areas in the weave pattern) and wet-out (actual wetting of the individual glass filaments). Woven roving with its open weave pattern provides better air removal during impregnation with the resin, but increases the possibility of resin-rich unreinforced areas in the laminate. Sizings available for this type of fabric are designed to be compatible with various resin systems and normally do not require subsequent finishing (surface treatment).

d. **Spun-strand roving fabric**, a more recent fabric development, does not offer as much bidirectional strength as does continuous-strand roving fabric. It does, however, provide multidirectional strengths, better wet-out, and improved layer-to-layer bonding characteristics (interlaminar shear).

Tapes or "narrow glass fabrics" are fabrics of limited width ranging from about ¼" to 8" and are woven from continuous-filament and staple fiber yarns. As in cloth fabrics, the tape is supplied with a selvedged edge. Thicknesses range from 0.003" to 0.025" with a range of breaking strengths (depending on width) of 35 pounds to 4700 pounds. Weaves are available in medium and light styles.

GLASS FABRIC WEAVES

The number of styles or weaves of fabrics being used by industry today is approaching 100. There are three basic weaves: plain, twill, and satin.

Since many variations of the basic weaves are available, the manufacturer, by selecting construction and style, can build into the fabric specific desired properties.

Below are descriptions of some of the more common weaves used in industry.

Plain Weave
The most common weave is the plain weave (sometimes called taffeta) in which one warp end (lengthwise yarn) passes over and then under one filling pick (crosswise yarn). This is said to be a "square" weave since the number of ends and picks are equal if the warp and weft yarns are of the same count. Plain weaves have these characteristics:

1. Firmest and most stable of the commercial weaves.
2. Afford good porosity with minimum yarn slippage.
3. Uniform strength pattern in same plane.
4. Afford ease of air removal in hand lay-up or molding processes.

Twill Weave
In these fabrics weft yarns pass over one warp yarn and then under more than one warp yarn. This construction results in a characteristic diagonal line pattern which appears on the surface. Crowfoot satin is a type of twill weave in which one warp end weaves over three and under one filling pick. Generally, twill weaves possess these characteristics:

1. More pliable than plain weaves.
2. More drapable since they will conform closely to complex or compound surfaces.
3. Permit weaving of higher counts than plain weaves.

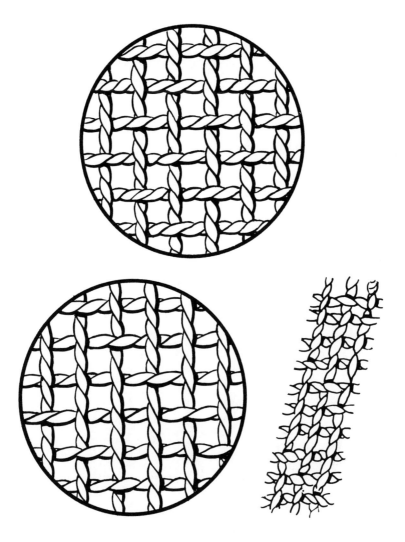

Long-Shaft Satin

This weave has one warp end weaving over four or more and under one filling pick. An 8-Harness satin weave (one warp end weaving over 7 fills and under one and staggered) is basically a long-shaft satin weave. These weaves have the following characteristics:

1. Most pliable of all of the weaves offering the maximum in drapability by readily conforming to compound curves.

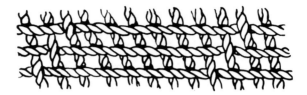

2. Permit production of laminates with high strength because of longer floats or unbent strands.
3. Can be woven in the higher construction counts or count density.
4. Generally less open than other weaves.

Basket Weave

This weave is similar to a plain weave but has two or more warp ends weaving as one end over and under two or more filling picks weaving as one pick. These weaves are typically: less stable than plain weaves, more pliable than plain weaves, and flatter and stronger than equivalent weight and count of plain weaves.

Unidirectional Weave

The unidirectional, or high modulus, weave is an adaptation of a basic textile weave. It employs a relatively high proportion of strong-warp yarns to weak-fill yarns to give this type of reinforcing fabric maximum strength in the warp direction. Generally, unidirectional weaves have these characteristics: maximum strength in one direction and impart high impact resistance to the laminate.

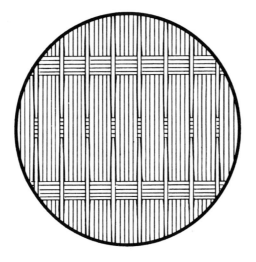

Leno Weave

The leno weave is produced by interlocking two or more parallel warp ends. The mock-leno is a modification of the simple plain weave in which two or more warp ends and two or more filling picks are closely spaced or bunched together, the closely spaced groups alternating with loosely spaced yarn. This style of weave produces a fabric having a rough texture, increased thickness, and additional porosity. Typical characteristics are:

1. Excellent mechanical bonding to the resin system.
2. Increased thickness at low fabric cost.
3. Pattern stability provided rather than drapability.
4. Provides for low number of count per inch which is desired in surfacing fabrics.

Following is a chart summarizing the common types of glass reinforcement.

Summary of Various Types of Reinforcements

Type or Structure	Form	Precursor	Weights Range	Dimensions Range	Cost Factor*	End-Use Application
Continuous Strand	Yarn, Continuous roving, Spun roving	Twisted single ends, Untwisted multi-strands, Looped single ends	176,400 yds./lb. to 200 yds./lb.	0.0026" diam. to 0.055" diam.	1–2½	Filament winding; production of other reinforcements; pultrusion products; preforms
Cloth Fabrics	Various styles of weave	Yarn—continuous	1 oz. to 40 oz./sq. yd.	0.0010 to 0.045"	3½–6½	Laminates
Woven Roving Fabrics	Various styles of weave	Continuous roving, Spun-strand roving, Staple yarn	15 oz. to 27 oz./sq. yd.	0.027 to 0.048" thick	2–3½	Laminates
Chopped Strands	Various lengths from ¼" to over 3"	Continuous-filament strands or spun strands	Not applicable	Various lengths	1–1½	Preforms and molding compounds
Reinforcing Mats	Chopped strand or Continuous swirl	Chopped strands or Continuous strands	¾ oz. to 3 oz./sq. ft., Mechanically bonded, 2 oz. to 10 oz./sq. ft.	Not representative unless compressed	1½–2½	Laminates
Surfacing Mats	Decorative overlay surfacing veils	Monofilaments	Generally less than 1 oz./sq. ft.	0.010 to 0.030" thick	5–6	Reinforcement of gel-coats; special surface effects in laminates

*Cost factor 1 indicates lowest cost, while 10 shows highest cost on $/lb. basis.

Arrangement and Amount of Reinforcements

As shown in the figure below, strength of the finished laminate increases in direct relation to the amount of glass.

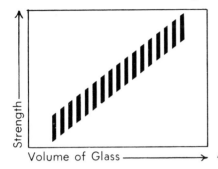

As glass content increases, strength increases.

RELATIVE COSTS OF GLASS REINFORCEMENTS
FORMS OF FIBER GLASS

Glass Form	Maximum Glass Content (Percent of weight)	Representative Cost (cents/lb.)[a] 30 60 90 120 150 180 210
Continuous roving	90	
Yarn	90	
Continuous swirl strand mat	50	
Spun roving	50	
Fabric (181 series)	75	
Fabric (unidirectional)	75	
Fabric (1000 series)	65	
Woven roving	60	
Woven spun roving	55	
Fabmat[b]	49	
Chopped strand mat	50	
Mechanically needled mat	50	
Chopped strands	50	
Chopped spun roving	50	
Surfacing and overlay mat	20	
Milled fibers	15	

1 2 3 4 5 6 7

a. Approximate delivered price, truckload quantities.
b. Trademark of Fiber Glass Industries.

A part containing 80 percent glass and 20 percent resin is almost four times stronger than a part containing opposite proportions of these two materials.

The arrangement of the glass reinforcing is of equal importance with respect to strength of the finished product.

Consider three cases: (1) when all glass strands are laid parallel (unidirectional); (2) when half of the strands are laid at right angles to the other half (bidirectional); and (3) when the strands are arranged in a random manner (isotropic).

DIRECTIONAL STRENGTH PATTERNS

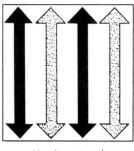  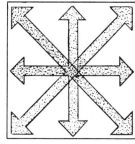

Unidirectional   Bidirectional   Multidirectional
(Isotropic)

When all the strands are laid parallel to each other, maximum strength results in one direction. This strength is supplied for end uses such as solid rods or solid bars. Product application would include guy strain insulators, golf clubs, and fishing rods. Strength perpendicular to the axis of the fibers is dependent on the shear strength of the resin.

When half the strands are laid at right angles to the other half, strength is highest in those two directions. Strength in any one direction is less than with parallel arrangement. This pattern of glass reinforcement is used in structural shapes.

When glass fibers are arranged in a random manner (isotropic), strength is equal in all directions. This arrangement is generally found in safety helmets, chairs, premix molded parts, luggage, and machine housings.

There is a relationship between the way glass is arranged and the amount of glass that can be loaded into a given object or product. The neater the arrangement or the more precise the placement, the greater the amount of reinforcement that can be placed.

In the case of an all-parallel arrangement, glass loadings from 45 percent to 90 percent can be obtained. The maximum theoretical amount is about 92 percent, while the practicable desired maximum is less than 80 percent.

When half the strands are placed at right angles to the other half, glass loadings range from 55 percent to 75 percent.

A random arrangement gives glass loading in a range of 15 percent to 50

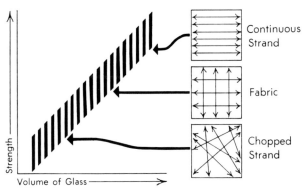

Continuous
Strand

Fabric

Chopped
Strand

Strength

Volume of Glass

Type of reinforcement determines maximum strength in a given direction.

percent. The relationship of amount of glass, strength characteristics, and arrangement of glass is shown in the figure below. Continuous parallel strands give the highest strength range, bidirectional arrangement the next highest, and random arrangement the lowest strength range.

Reinforcements are sold in forms which permit the designer or user to utilize this directionality to maximum advantage. The basic forms are continuous strand, fabric, woven roving, chopped strand, reinforcing mat, and surfacing mat.

Continuous strand gives unidirectional reinforcement. In filament windings it is utilized to give a balance of reinforcement in the desired directions and at the desired angle (see description of Filament Winding Process). Continuous strand is one of the lowest-cost reinforcements in the fiber glass field.

In summary:

1. Fabric essentially reinforces the object in two directions.
2. Woven roving gives high strength and is lower in cost than conventional glass fabrics.
3. Chopped strand gives random reinforcement.
4. Reinforcing mats are lower in cost than fabric and give random reinforcement.
5. Surfacing mats give virtually no reinforcement but permit a decorative and smoother surface finish.

As mentioned earlier, a definite relationship exists between reinforcement, molding methods, and engineering properties. In general, increased molding pressure (matched metal versus contact) decreases laminate thickness and increases glass percentage (with resultant lowering in resin content). Strength properties are improved as the quantity of glass reinforcement is increased.

Different types of fiber glass reinforcements, molded by the same method, vary widely in thickness and weight per ply of finished laminate. Differences in shop practices or molding techniques will also produce variations in laminate thickness and weight and strength properties for any particular type of reinforcement.

Owens-Corning Fiberglas Corporation sponsored a comprehensive test program in 1959 and 1960, the results of which were published in *Marine Design Manual for Fiberglass Reinforced Plastics,* by Gibbs and Cox, Inc. (McGraw-Hill Book Company, Inc., 1960). The information in Chapter 5 provides a more comprehensive discussion of "relationships between reinforcements, molding methods and properties" of fiber glass polyester laminates.

The test program was directed toward the fiber glass boat industry since the predominance of manufacture at that time was by methods of contact molding (hand lay-up). Since marine applications were of primary considera-tion, all mechanical properties were obtained from tests of laminates in the "wet" condition or immersed in water at room temperature for 30 days. The information developed was based on the use of over 200 test panels and over 6000 individual tests. Some of the tables summarizing these test results in Chapter 5 of *Marine Design* are: Table 5-2   Plies versus laminate thickness in inches; Table 5-3   Plies versus laminate weight; Table 5-4   Glass content versus specific gravity; and Table 5-5   Void content of contact molded polyester fiber glass laminates.

There are also a number of tables summarizing the data developed regard-ing mechanical properties.

A method for calculating void content or percent voids in a laminate has been developed (Military Specifications MIL-P-17549B (Ships), October 2, 1956) as follows:

$$\text{Percent voids} = 100 - 100\,a \left[\frac{d}{c} + \frac{e}{b} + \frac{f}{g} \right]$$

a = specific gravity of the laminate
b = specific gravity of fiberglass = 2.55
c = specific gravity of cured resin = 1.18 to 1.24 as obtained from
    manufacturers
d = resin content, by weight
e = glass content, by weight
f = filler content
g = specific gravity of filler

The above method is not exact because the assumption is made that the resin system has the same density in a laminate as it does in an unreinforced casting. The net result is a possible overstating of the void content.

A more precise determination of Void Content can be made by using the Standard Method of Test for Void Content of Reinforced Plastics (ASTM D 2734-70 Pt. 36, 1974 ed.)

Voids are generally the result of the entrapment of air during the con-struction of a laminate by hand lay-up methods. Any volatile components in the resin that might be released during cure will contribute to void content.

It is desirable to keep the void content at a low value, but it is not unusual to find void contents of 1 to 3 percent in laminates that have been prepared in a workmanlike manner.

SELECTING GLASS REINFORCEMENTS

The laminate designer is faced with selecting the appropriate reinforcement to do a job, and in all probability he will work toward a design which will provide the desired strength at the least expenditure for labor and materials. The following summary may assist him.

Woven Roving
Woven-roving reinforcements are more costly than mat but less costly than cloth fabrics.

Advantages

1. Good handling and drapability for use in hand lay-up.
2. Provide a thicker build-up per ply of laminate than cloth.
3. Provide a high glass content per ply.
4. Exhibit high directional physical strength and moduli for orientation in areas of high stress.
5. Offer good impact resistance.

Disadvantages

1. Difficult to wet-out or thoroughly impregnate (wet-through) with resin because of the tightly compacted filaments in the strands. Poor wet-out can lead to inadequate bond between the filaments that make up the roving.
2. Air bubbles are easily entrapped which results in increased void space and presents areas of attack by water and chemicals.
3. Lend themselves to greater nonuniformity or presence of resin-rich layers between the layers of rovings. Resin-rich layers are subject to crazing, cracking, poor strength, and poor interlaminar bond.
4. Strength properties are highly directional in each layer.

Mat
Two types of mat are considered in this discussion: bonded mats, in which the chopped strands are held together in desired position by a resin binder compatible with the resin system; and those made from random strands of glass held together mechanically by a so-called needling.

Advantages

1. Low cost per square foot per unit thickness.
2. Balanced physical properties in all directions.
3. Good interlaminar bond because of the interlocking of the fibers.
4. Good handling and drapability in forming into complex contours.

Disadvantages

1. Laminate thickness is somewhat difficult to control.
2. During molding the glass fibers will shift.
3. Presents a poor finished surface by itself, and therefore requires the addition of a surface veil or a resin-rich top coat.
4. Laminate will exhibit a lower glass content than cloth or roving which in turn results in a lower modulus of elasticity for equal thickness. This is overcome by use of thicker sections.

Cloth

Cloth is available in plain, satin, or unidirectional weaves, but is one of the more expensive types of glass reinforcement. Where applications demand high performance and structural efficiency in terms of strength-to-weight ratios, its relatively high cost can generally be justified.

Advantages

1. Effective surfacing material to cover woven rovings and mats for better appearance and increased strength.
2. Exhibits good workability to eliminate excess resin and provides a high glass content in laminate.
3. Provides good control of thickness and uniformity of glass content resulting in minimum deviation of physical properties.
4. Particularly suited for reinforcement in areas of high stress.

Disadvantages

1. In laminates utilizing heavy cloth plies, a resin-rich bond may develop (similar to woven rings) which will cause a weakness in interlaminar shear. This tendency will show up under edgewise compressive loadings and cause failure by delamination.
2. Labor costs may increase if thick sections are involved because of the number of layers required to obtain desired build-up of thickness.

OTHER REINFORCEMENTS

Reinforcing materials other than glass are briefly discussed here, as well as their uses and importance in the reinforced-plastics industry. Some of the standard texts listed in the appendix provide a more comprehensive discussion of each material if further details are required.

Asbestos

Short asbestos fibers, such as floats, have been used for many years in phenolic molding compounds and in vinyl floor tile, primarily to impart the desirable benefits of flow control, heat resistance, dimensional stability, and low cost.

In the past twelve years, producers of asbestos fibers have made available specially prepared grades for use as reinforcements for plastics. They are offered commercially as loose fibers, in papers, mats and felts; and each manufacturer provides a variety of technical information on their appropriate use with plastics.

One of the growing commercial uses is in corrosion-resistant applications, such as reinforcing resin-rich liners on pipe interiors. Asbestos fibers have demonstrated their effectiveness in special military and electrical applications, particularly where premix molding compounds are used.

Metal, Wood, Vegetable, Papers, and Inorganic Materials

Most of the reinforcements in this group are of specialized nature and at the present time are not used in significant quantities in the reinforced-plastics industry, with the exception of sisal, which is used in considerable volume in low-cost "gunk" molding processes. Standard texts listed in the appendix provide a comprehensive discussion.

Natural or Synthetic Organic Fibers

Some of these fibers, particularly the synthetic organic types, are becoming more important in the reinforced-plastics industry.

Dynel, a trademark of Union Carbide Corporation, is a synthetic fiber produced by extruding an acetone solution of acrylonitrile vinyl chloride copolymer. It is resistant to strong acids and extremely resistant to alkalies and many organic solvents. It has a comparatively low softening point of 250°F and is classed as self-extinguishing. It is available in the form of fabrics and surfacing veils.

Orlon, a trademark of E.I. duPont deNemours & Co., Inc., is a synthetic fiber consisting primarily of acrylonitrile and is highly inert to strong acids, sunlight, and the effects of weathering. It is generally resistant to moisture, organic solvents, and weak alkalies, and shows good resistance to mildewing and bacterial attack. It is comparatively strong, dimensionally stable, and slow-burning, with a softening point of 455°F. It is available in woven fabric of spun (continuous) roving.

Dacron, a trademark of E.I. duPont de Nemours & Co., Inc., and **Terylene,** a trademark of Imperial Chemical Industries, are linear polyester fibers and show toughness, strength, and abrasion resistance. Chemically they are resistant to weak acids, alkalies, and organic solvents, have a softening point of 450°F, and are slow burning. They are available as rovings, nonwoven fabrics, or thin mats made from untwisted continuous yarn and as woven and knitted fabrics when produced from continuous yarns.

Encrow, a trademark of American Enka Company, is a reinforcement manufactured from a saturated polyester thermoplastic. It has been produced as a woven roving at a weight of 25 ounces per yard squared. Fabric at four ounces per yard squared has been produced. Recommended use is with glass fiber reinforcement to improve abrasion and corrosion resistance of the composite.

Fabrics and continuous strands made from **polypropylene** have been used in the reinforced-plastics industry. They are light in weight and exhibit good corrosion resistance.

Cotton fiber possesses a natural twist which is beneficial in contributing added strength to yarn and consequently to woven fabrics. Cotton is durable, absorbent, resistant to heat and to stretching; it has low resistance to acids and mildew, but good resistance to alkalies and solvents. Fabrics made from cotton fibers, twisted and plied, are available in weights of 2 to 10 ounces per square yard, mostly in conventional weaves.

7

Molding Systems

Much of the discussion in this book relates to a step-by-step directed combination of a reinforcement and a thermosetting resin to produce a composite material or laminate. For example, in the process of hand lay-up a step-by-step procedure results in a completed laminate that represents the sum of the parts.

Developments in molding compounds, SMC (sheet-molding compound), and BMC (bulk-molding compound) confirm that a system concept can be used successfully for many proven and accepted product applications and lead the way to many new applications.

This system concept might be compared, liberally, to vacuum forming in the thermoplastic industry or to sheet stamping in the metal industry.

Complete details on the production of the one-component molding system are beyond the scope of this book. Many of the formulations are proprietary to the developer, but the general outlines of the developments can be traced.

The technical breakthrough has been the development of thickening systems and profile improvement additives which are compatible with all of the other components and which also permit a reasonably long shelf life for the product. The desired thickening system exhibits viscosity characteristics that provide good flow of both the glass reinforcement and resin as a homogeneous material into all parts of the mold.

These one-component molding systems combine resin thickening agents, extenders, catalysts, pigments, glass fibers, and release agents and offer to the molder a transportable starting material that can be handled with a greater degree of automation and less waste than was possible with earlier conventional molding techniques. By eliminating inefficient individual component handling problems of more conventional FRP molding, SMC and BMC can lay claim to increased production rates per dollar of tooling and mold machine investment. The matched metal molding fabricator welcomes this improvement.

Since about 1969, many automotive components have been molding from so-called low profile SMC, a system which yields an especially smooth surface.

Selected components of the 1974 Corvette are produced using sheet molding compound (SMC). (Chevrolet Motor Division, General Motors Corporation)

Grill opening panel for 1974 Mustang II produced from sheet molding compound by Goodyear Aerospace Corporation.

Electrical contactor/starter by Arrow-Hart, Inc.

These included integrated front end panels, headlamp housings, spoilers, window frames, air deflectors, and rear wheel opening covers. Other grades of SMC are molded into appliance housings, computer housings, welding helmets, swimming pool filter bodies, tote boxes, battery boxes, laundry tubs, bath tubs, septic units, and shutters. Electrical applications include street light housings, circuit breaker bases, fuse holders, and switch gear parts. Some of these applications would utilize the BMC system.

Let us define and characterize the sheet-molding compound. SMC is a machine-made product. Producing machines are semi-automated and thus require minimal supervision. The product is generally made in 4-foot widths and in continuous lengths of transportable quantity. The process involves a directed combination of reinforcement and polyester resin together with catalysts, pigments, extenders, thickeners, and other additives. The reinforcement, generally continuous-strand roving, is chopped into desired lengths and comes to rest on a moving film of resin-covered polyethylene which serves as the carrier to make it transportable. Special formulated resin systems are layered onto the chopped reinforcement, then an overlay of another sheet of polyethylene film is applied. This sandwiched SMC is fed through rollers to distribute the resin, then onto rolls of desired lengths. Glass content can be controlled in the range of 25 percent to 45 percent by weight to meet the particular requirements of the end-use application. Production of low-shrink or low-profile SMC would dictate a change in the resin system used.

Bulk-molding compound, BMC, is a product that previously might have been termed *premix,* or *gunk.* Present-day BMC utilizes the same thickening principle used in SMC and can be described as an admixture of polyester resin, reinforcements, fillers, catalysts, pigment, and other additives. BMC is supplied to the molding operation in a transportable bulk form. Generally speaking, BMC contains a relatively high amount of filling materials such as clays, talcs, calcium carbonates, and asbestos. Glass content, chopped from continuous-strand roving, can be controlled in the range of 15 percent to 25 percent. The length of the reinforcement in BMC is in the range of $\frac{1}{2}''$, whereas in SMC longer length fibers are used. The mechanical properties of BMC are generally lower than those of SMC. This is not incompatible since the end-use application generally influences the desired properties of the material that will be molded.

SMC and BMC are available from a number of suppliers. Industries requiring large in-house quantities have set up machinery and equipment to produce compound on a continuous basis, and purchase the reinforcement as well as the resin system in the quantity to support their needs.

Considerable research and development effort is being put forth by polyester resin producers who want to sell to this market.

These companies who produce and market SMC and BMC systems have taken on an added responsibility which is probably most welcome to the purchasing agent. They offer customers assistance with molding production problems. Before the development of the one-system concept, users were uncertain where to find help with production problems, although the resin supplier probably was contacted the majority of the time.

The market growth potential for the one component system is promising. It is expected that considerable research and development effort will be continued. Much of the market growth, typical of the RP industry, is not created

The sculptured look that typifies the appearance of sports cars in the world of snowmobiles is usually based on reinforced plastics. Sheet molding compound (SMC) is used for the production of hoods, hood decks and instrument consoles. This produces smooth-surface parts with molded-in color and tough performance characteristics, and provides great design flexibility. (Goodyear Aerospace Corp., Commercial Plastics Div., Akron, Ohio)

The International Harvester, Cadet 76, lawn tractor utilizes a fiber glass reinforced plastic hood and grill molded from SMC by Goodyear Aerospace's Commercial Plastics Division.

The integrally designed 1973 GMC motor home manufactured and marketed by GMC Truck and Coach Division of General Motors Corporation features extensive use of fiber glass reinforced plastic body components made from compression molded sheet molding compound.

but represents substitution or replacement of FRP for more conventional materials. Still, FRP market growth presents the usual supplier-customer considerations:

1. The product from the one-component system must meet the specified physical properties of the purchaser, these properties having been established by the use of more conventional materials.
2. The market generally requires the end product to have a decorative finish or protective coating—in other words, the system should be compatible with standardized painting and decorating systems.
3. Complete understanding between the customer and the supplier is necessary to minimize and utilize effectively the time required for qualification.

The dynamic FRP industry has the capability of meeting the challenge.

8

Design Criteria

Rules and criteria for designing with steel, aluminum, and wood have been available to engineers for a number of years, but only in recent years has similar engineering information been available to the designers of laminates and structural shapes of reinforced plastics.

Principles used in structural design with reinforced plastics can be compared in many ways to designing with steel or aluminum. Fabrication of fiber glass laminates likewise can be compared in many ways to working with wood with differences in cutting and fastening.

In the normal design of structures using steel or aluminum within the elastic range, the assumption is made that the steel or aluminum is never stressed beyond the proportional limit. Within this elastic range the material is said to obey Hooke's Law of Proportionality; i.e., the unit stress is directly proportional to the unit strain which means that the material will return to its original shape when the load is removed.

Glass-reinforced laminates are complex materials and do not always behave according to Hooke's Law; but in the majority of cases, this assumption can be made without appreciable error in the results. This becomes more obvious particularly when one recognizes that the use of recommended Design Factors (Factors of Safety) restrict allowable loads to a limited area of the stress-strain relationship.

Therefore, in design procedures and theories for laminates composed of fiber glass and resin combinations, as well as for structures fabricated from these laminates, the following assumptions are valid: The combination of fiber glass and resin acts as a composite, and the combination is considered elastic and obeys Hooke's Law.

Typical stress-strain curves for fiber glass laminates follow.

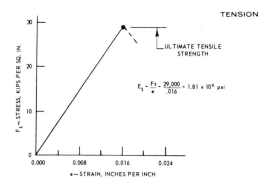

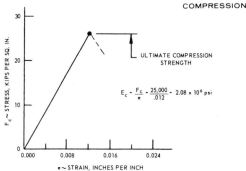

From the stress-strain curves it is readily apparent that the stress-strain relationships for reinforced plastics are generally similar to those of most structural materials insofar as they exhibit a linear portion followed by a nonlinear portion. Like wood and some metals, no yield point exists.

It is important to emphasize that for stresses applied in the direction of the reinforcement, the deviation of the upper nonlinear portion of the curve from the lower linear portion is usually small at the point of failure. This limited deviation of the curve indicates that strain deformations are quite small and that the material has low ductility. This low ductility of glass fiber reinforced plastic does not allow for stress relief in areas of stress concentration such as notches, holes, reduction in area, and sharp angles, when compared to a more ductile material.

Design Factors

In any design problem involving the use of fiber glass laminates in fabricating load-bearing structures, it is important to consider design factors. Design factor is defined as the ratio of the ultimate strength of the material to the allowable working stress.

$$D.F. = \frac{\text{Ultimate Strength}}{\text{Allowable Working Stress}}$$

In many fields the allowable working stresses or design stresses are specified by codes or recognized authorities. In more conventional materials of construction—such as steel—the values for design factors (factors of safety) are backed up by many years of experience. This means in the case of steel, as well as many other metals, that the values for ultimate strength as well as other mechanical properties have been established with a high degree of confidence.

Reinforced plastics, a relatively new material of construction, does not have a comparable wealth of experience and demonstrated performance.

The basis for the establishment of mechanical properties should be considered. Laboratory work has shown that using conventional strength-of-material formulas and the results of selected bending, tensile, and compressive tests of entire cross sections that typical mechanical property values may vary significantly from values obtained from testing small coupons in accordance with recognized test methods.

Metals are considered to be homogeneous and can be expected to show equal strengths in all directions. The degree of homogeneity in reinforced plastics, however, is a function of the manufacturing technique and the orientation of the reinforcement. The directional strengths in fiber glass composites are a function of the type and amount of reinforcement. Since stress patterns are a function of the homogeneity of a material, this variable must be taken into account in establishing a design factor. The low ductility of glass-reinforced plastics does not accommodate stress relief in areas of stress concentration. This means that the stress analysis formulas used with metals may not always be applied similiarly to reinforced plastics.

As with all structural materials, damage or deterioration caused by mechanical abrasion or unusual handling conditions cannot be accommodated by increased design factors. Such conditions must be eliminated if reinforced plastics are to be used successfully.

Strength properties of reinforced plastics are reduced when exposed to elevated temperatures. The amount of change will vary with different resin-glass systems. Changes in the values of the Modulus of Elasticity will also vary with different resin-glass systems. The proportional reduction in physical properties for a typical system with respect to temperature is shown below:

Temperature	Ultimate Tensile or Compressive Strength	Modulus of Elasticity
75°F.	20 Kips/in^2	2.3 × 10^6 psi
125°F.	16 Kips/in^2	1.8 × 10^6 psi
175°F.	12 Kips/in^2	1.4 × 10^6 psi
200°F.	10 Kips/in^2	1.2 × 10^6 psi

Minimum design factors of 4.0 are suggested for reinforced plastics for structural applications. Responsible manufacturers use a minimum design factor of 5.0 for filament-wound fiber glass pipe. Obviously, a piping system conducting fluids is not a structure in the true sense of the word, but the careful considerations of the ratio of the ultimate strength to the allowable working stress does apply.

The preceding discussion is fundamental to the establishment of design factors since many of the considerations presented can influence the values selected for ultimate strength. Once the designer has established the ultimate strength values that can be used, he can proceed with the consideration of the following in the selection of a design factor:

1. Accuracy of the estimated loads on the structure.
2. Precision of analysis and stress determination: Where the analysis for stress is precise and accurate a smaller factor may be used; inexact or approximate analysis requires a greater factor.
3. Deterioration of strength properties due to environmental conditions: The designer should consider, in addition to the effect of temperature, the effect of unusual wind loadings, as well as the effect of wind gusting. The effect of seismic loading should also be considered.
4. Nature of loading: Certain types of load have greater effects than others. The effect produced by the different types of loading on the ultimate strength of the material should be known so the most appropriate factor can be selected. The following minimum factors for various load conditions should be considered for fiber glass structures:

Type of Loading	*Minimum Design Factor*
Static short-term loads	2.0
Static long-term loads	4.0
Variable or changing loads	4.0
Repeated loads-load reversal or fatigue loading	6.0
Impact loads	10.0

Design factors are dependent on many variables which only the design engineer can analyze. The factor should be selected only after all uncertainties have been thoroughly considered—including the assumptions involved and the theories used. Every problem presents its own peculiarities and requirements and, therefore, the judgment and experience of the designer plays an important part.

The final selection of a design factor (factor of safety), unless established by some other authority, becomes the responsibility of the designer.

9

Mechanical, Electrical, and Physical Properties

Mechanical, electrical, and physical properties are important to the user of reinforced plastics for a number of reasons. They can be used:

1. To define properties that may be important to the engineer in developing a design.
2. To compare one material with another.
3. To serve as a basis for quality control.
4. To guide the user to proper end-use applications.

It is important to remember that direction of reinforcement will affect some of the mechanical properties or that properties will vary in different directions in the same plane of the laminate under test.

Some of the more important properties are discussed below.

FLEXURAL STRENGTH

This property is a measure of bending strength, and defines how great a non-moving load can be applied before the test specimen yields or breaks. Units are normally reported in thousands of pounds per square inch (10^3 psi). The higher the value, the greater the flexural strength or resistance to bending, when a load is applied. Size of specimen is usually 6 inches by $1''$ x $1/8''$ loaded over a 4-inch span. Flexural strength values are reported at room temperature and, generally speaking, flexural strength values for reinforced plastics decrease as the temperature increases.

FLEXURAL MODULUS

This property is associated with the stiffness of materials and is used to calculate how far a material will bend when a bending load is applied to it. Values are

How heavy a load can it stand?
How far will it bend?

Load

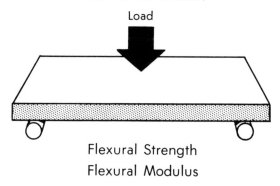

Flexural Strength
Flexural Modulus

generally reported in millions of pounds per square inch (10^6 psi). Higher values for a material mean that the material is more resistant to bending (measured as deflection) when compared to another material under comparable conditions. Values for flexural modulus can be calculated from information developed in the determination of flexural strength.

TENSILE STRENGTH

The value for this property indicates how large a nonmoving load the material can withstand before breaking due to elongation when tested in accordance with stated test methods. Units are normally reported in thousands of pounds per square inch (10^3 psi). Higher values indicate materials which can withstand a greater pull prior to breaking. Test specimens are usually machined to a special shape for use in this test.

How big a pull can it stand?
How much longer will it get?

Pull

Pull

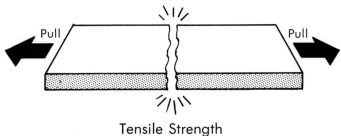

Tensile Strength
Tensile Modulus

TENSILE MODULUS

A test specimen subjected to tensile forces will elongate. The ratio of tensile force to elongation is known as tensile modulus (also called Young's modulus), and this value can be used to calculate how much longer a material will get when a predetermined load is applied. Units are normally reported in millions of pounds per square inch (10^6 psi). Higher numbers indicate materials that will exhibit less elongation than other materials tested under like conditions.

Generally speaking, the higher the modulus, the more rigid or more resistant the material is to stretch. Values for calculating modulus are available from the conduct of the Tensile Strength Test.

COMPRESSIVE STRENGTH

This value indicates to what extent a material can take a nonmoving load before it is crushed by the force that is being applied to it. Units are normally reported in thousands of pounds per square inch (10^3 psi). Higher values indicate an ability to take a correspondingly higher load. The engineer should remember that in reinforced plastics, the compressive strength values will vary with the direction of reinforcement and that three separate values can exist for the same material because of the three different directions in which the reinforcement can be oriented.

How big a push can it stand ?

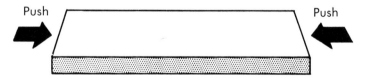

Compressive Strength

ELONGATION

As discussed previously, a material will increase in length when subjected to pulling forces. Elongation values tell how much longer a material will become before it breaks. In each of the previous discussions, the behavior of the material has been stated to be under conditions of nonmoving loads. Under this type of loading there is a fundamental difference between reinforced plastics and some of the common metals. This can best be shown by the illustration of a typical stress-strain curve where stress of load is plotted as the ordinate and strain or elongation is shown by the abscissa.

For most common metals there is a yield point in the stress-strain curve. At the yield point the increase in elongation as compared to the increase in load is significantly different from the ratio initially exhibited. The metal will eventually break.

In the case of reinforced plastics, however, there is no definite yield point and the curve is essentially a straight line until the material breaks. In metal design practice, the basic reference value is yield strength. In reinforced plastics the value for ultimate strength is the valid number to use.

Units of elongation are reported in percentiles (%). Higher numbers indicate materials which will stretch or elongate further before they yield or break.

The Modulus of Elasticity in tension is the number resulting from dividing the stress (in pounds) by the elongation (in inches).

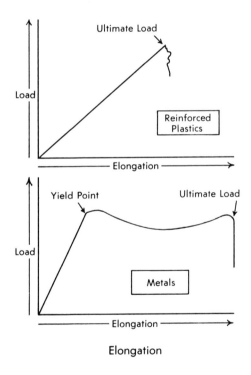

Elongation

IMPACT STRENGTH

When a moving load strikes a stationary object, the effect is called "impact." The impact strength of a material is a measure of how much energy is absorbed by a test specimen when the amount of load (or work) is sufficient to break the material.

Many different tests have been developed to measure impact. In current use is the Izod impact test, which, except as a nonsophisticated screening method, is more applicable to homogeneous materials than it is to composites. The results of the test indicate the energy absorbed at failure and not at partial failure, as usually occurs in laminates made from reinforced plastics. Units are reported in foot pounds per inch of width of notch. Higher values indicate that the material will absorb more energy before it is broken with a moving weight.

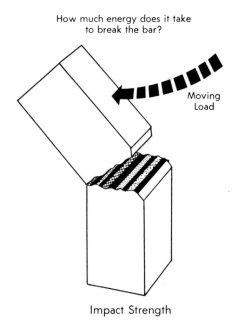

How much energy does it take
to break the bar?

Moving
Load

Impact Strength

THERMAL CONDUCTIVITY

This property is commonly called the "k" factor and represents the measure of heat transfer by conduction. It is reported as the number of British Thermal Units (BTU) transferred per hour per unit area (1 sq. ft.) for a thickness of one inch, when a temperature difference of one degree Fahrenheit exists between the two surfaces of the plate ($BTU/hr/ft^2/°F/inch$).

Higher values indicate that more heat is conducted through the material and that it is not as good an insulator as a material having a lower value.

SPECIFIC HEAT (THERMAL CAPACITY)

This property defines the amount of heat required to raise the temperature of one pound of material one degree Fahrenheit. Units are reported as BTUs per pound per degree Fahrenheit (BTU/lb/°F).

Higher values indicate that the material requires a greater amount of input heat energy to raise the temperature of the material. Water by definition has a specific heat of 1.0.

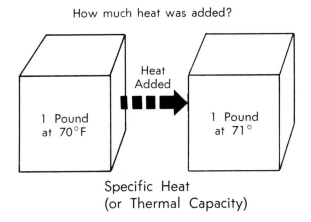

How much heat was added?

Specific Heat
(or Thermal Capacity)

FLAMMABILITY

This property describes how fast a combustible material will burn when subjected to a recognized flammability test. In the ASTM test, D–635, a flame is applied to one end of a strip of material. When the material starts burning, the flame is removed and the time to consume a given amount of material is measured. Units are inches per minute (in/min). Higher values would indicate that the material would burn at a greater rate.

When materials are classified as self-extinguishing (S.E.), the test specimen would not continue to support combustion once the flame is removed. This test does not truly represent the conditions that might be experienced in an industrial fire and its results are open to interpretation. There are a number of other "flammability" tests in current use, some of which have been discussed earlier.

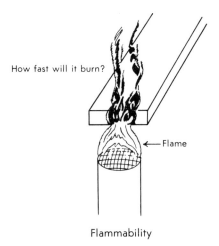

How fast will it burn?

←— Flame

Flammability

ROCKWELL HARDNESS

A material's resistance to indentation or penetration by another substance is Rockwell hardness. The Rockwell method for measuring hardness forces a steel point into the material and then measures the penetration of the point. Units are reported as Rockwell units with the appropriate suffix letter, which describes the shape of the point and the load applied during the test.

A higher number with the same suffix letter indicates a harder material which is more resistant to penetration.

BARCOL HARDNESS

Barcol Hardness as determined by a Barcol Impressor is probably more common to the reinforced-plastics industry. It is a comparable measure of hardness using a scale of values of 0 to 100, and the higher the number, the harder the material. A Barcol Impressor is used on occasion to measure and follow the progress of cure in reinforced plastics. Fillers and reinforcement may cause considerable local variations in the readings.

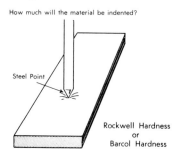

How much will the material be indented?

Steel Point

Rockwell Hardness
or
Barcol Hardness

DIELECTRIC STRENGTH

Measure of this electrical property give an indication of the ability of a material to act as an electrical insulator. Values report how great an electrical voltage can be built up on one face of a material before it is communicated to the other face. Units are reported as volts per mil of thickness (volts/mil).

In the evaluation of reinforced plastics for selected end-use applications, the value of specific inductive capacitance (SIC) is of importance to electrical engineers. A more common name for SIC is "dielectric constant."

The dielectric constant of a material is the ratio of the capacitance formed by two plates with the test material between them to the capacitance of the two plates separated by the same distance in a vacuum.

Another electrical property of importance is that known as "loss tangent," which is defined as the ratio of the conduction current to the displacement current when the material under test is placed between two plates. The procedure is similar to that used in determining the value of the dielectric constant. "Loss tangent" is also known as "dissipation factor."

On occasion a third property, known as "loss factor," may be of interest. It is defined as the product of the dielectric constant and the loss tangent.

Arc Resistance, as measured by ASTM D-495, is another electrical property of import to end-users. The units of measurement are seconds and the higher the value reported, the better the resistance of the material to breakdown by arcing.

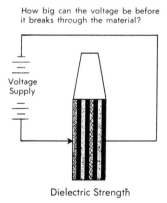

How big can the voltage be before it breaks through the material?

Voltage Supply

Dielectric Strength

SPECIFIC GRAVITY

Specific gravity is the ratio of the weight of a material to the weight of an equal volume of water.

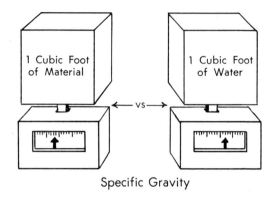

Specific Gravity

DENSITY

Density is defined as the weight of a material per unit volume. Units are generally reported in pounds per cubic inch (lbs/in^3).

DEFLECTION TEMPERATURE

Formerly called "Heat Distortion Temperature," this is a measurement of the temperature at which a specified load will cause a material to bend or deflect a specified amount. In the conduct of the test, a load is applied in bending to cause 264 psi of stress in the material. The temperature is increased at a set rate until the material bends 1/10″ at the center, at which time the temperature is recorded and reported.

Units are reported in degrees Fahrenheit. Higher values mean that a material can be heated to a higher temperature before it deflects 1/10″ under the arbitrary load of 264 psi.

CONTINUOUS HEAT RESISTANCE

This is an empirical test which indicates the maximum temperature the material should be subjected to in a continuous application.

Weight loss is closely related to this property and is a reporting of percentage weight loss after continuous exposure to a stated temperature for a predetermined number of hours. It, too, is used to estimate the serviceable life of a material at elevated temperatures.

THERMAL COEFFICIENT OF EXPANSION

Tested values of this property measure the amount of expansion or contraction that will occur in a material when it is heated or cooled. It is generally reported in units of inches per inch of original length per degree Fahrenheit change in temperature. Higher values indicate a greater change in length per unit of temperature.

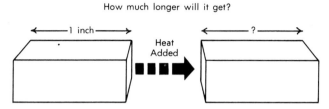

How much longer will it get?

Thermal Coefficient of Expansion

STRENGTH-TO-WEIGHT RATIO

This is a "property" which is the ratio of the tensile strength to the density. For example, a reinforced plastic exhibiting a tensile strength of 100,000 psi at a density of 0.072 pounds per cubic inch would have a ratio of $\frac{100,000}{0.072} = 1.39$ x 10^6. Steel would have a ratio of $\frac{240,000}{0.283} = 0.85$ x 10^6. The values obtained are empirical and sometimes serve as a factor of evaluation.

Below is an index of the recognized test methods for plastics, together with the test number applicable to Federal Specifications 6P–4066 and the equivalent ASTM method.

Test Methods For Plastics

Property	Federal Spec. 6P-4066	ASTM
Flexural Strength	1031	D790
Flexural Modulus	1031	D790
Tensile Strength	1011	D638/D651
Tensile Modulus	1011	D638
Compressive Strength	1021	D695
Elongation	1011	D638
Izod Impact	1071	D256
Thermal Conductivity	N.T.	C177
Specific Heat	N.T.	N.T.
Flammability	2021	D635
Rockwell Hardness	1081	D785
Dielectric Strength	4031	D149
Specific Gravity	5011/5012	D792
Density	5011/5012	D792
Deflection Temperature	2011	D648
Continuous Heat Resistance	N.T.	N.T.
Thermal Coefficient of Expansion	2031	D696

N.T. means no standardized method from these sources.

So far some of the theoretical aspects of glass-reinforced plastics have been discussed, including their advantages and disadvantages, resin systems used, reinforcements and their use, design criteria, and mechanical, electrical, and physical properties. In Part Two, the practical application of this information as it relates to the use of reinforced plastics in industry is discussed.

Part Two
PRACTICES

10

Selecting a Supplier

There is much to consider in selecting a supplier of a product, particularly one as new as reinforced plastics. The RP industry too has had its share of "garage shop operators." There are two kinds of "garage shop operators," good ones and bad ones. The consumer is well advised to make certain he deals with the former.

The consumer who buys reinforced-plastic structural shapes or a piping system is not purchasing the individual specific item. Although it is true that the product of a successful manufacturer must be competitive in terms of price and quality, the buyer has the right to expect the backup of a top-flight service organization as well.

Some companies are all talk and "no action," and this quality is easily detected by the seasoned buyer. Technical depth as well as financial responsibility is important to back up the product. This depth should extend throughout the manufacturer's organization right down to the last man who puts the final touch on the installation.

There is considerable merit in choosing a supplier whose products are made by machine methods where the process is continually monitored by automatic controls. In addition to eliminating many of the human element factors, the manufacturer offering products made by automated methods can supply products with a far higher level of reproducibility than the job shop utilizing hand lay-up techniques.

Once the purchaser has some experience with the performance of the specific product in service, he will normally draw on this experience with respect to future applications. The purchaser has every right to expect that the supplier will provide additional material that is comparable to the previous order with respect to performance and reproducibility. Generally speaking, the job-shop supplier is at a disadvantage where reproducibility is a factor.

Resin suppliers can provide a list of responsible manufacturers or custom fabricators. Fortunately this list is growing.

In addition to financial responsibility, the ability to deliver on schedule is important to the purchaser. It is very likely that the availability feature compared to that of conventional materials may have been the compelling reason for a purchaser to have decided in favor of reinforced plastics.

Some doubt is cast on the acceptability of a supplier if he is reluctant to accept reasonable purchase specifications for a reinforced-plastic product.

If the job involves custom fabrication to the drawings of the customer, then a visit to the fabricator's plant may be well worthwhile. If the job is a proprietary product, such as sheet, tube, pipe, tanks, structural shapes, etc., then in most cases a visit is a waste of time. In a competitive industry such as this, technical know-how often makes the difference between success and failure among manufacturers, and it would be less than forthright to indicate that such technology is not zealously guarded.

The first order should be placed with a supplier whose offering has been evaluated by:

1. Reputation and performance in the industry.
2. Level of reproducibility of product.
3. Willingness to build to reasonable performance specifications.
4. Willingness to translate his experience to the design problems of the new customer.
5. Adequate facilities to meet required delivery schedules.
6. Financial responsibility and a willingness to back up his recommendations.
7. Technical depth with technical service.

The first job should not be bought on price alone.

11

Specifications for Purchase

Technology in reinforced plastics has progressed far in the relatively few years since this material was first used as a replacement material for more conventional materials.

Literature reports that early in 1940 a laboratory technician accidentally spilled a catalyzed polyester resin over several layers of glass cloth. The resin cured overnight. Later examination pointed to the superior mechanical strength of this composite—so the reinforced plastics industry was born.

It is a matter of record that a goodly number of years passed before acceptance of test methods and specifications for describing reinforced plastics came into being. Everyone in the industry should be willing to admit that additional tests (nondestructive variety), additional specifications, and additional standards of performance are still needed. Workable and practicable specifications are needed in the purchase of glass-reinforced materials today; and congruently, the adherence by the manufacturer to reasonable specifications so that his product will perform adequately is a must if this industry is to continue to grow and to be accepted by users.

As with any new industry, poor performance brought about by inferior materials and workmanship has sometimes plagued the reinforced plastics industry. Use of specifications and proof of their compliance judiciously applied and directed to obtaining optimum performance can contribute to the continued growth of reinforced plastics.

Three factors important to optimum performance in a particular end-use application are:

1. Good specifications for use by the supplier and/or custom fabricator.
2. Consistency of product within itself plus reproducibility from piece to piece.
3. Proper and thorough inspection.

The plant or operation planning to utilize reinforced plastics should establish specifications for items it will purchase, as well as specifications for structures it may fabricate.

These might be stated as follows:

In-House Fabrication	Outside Purchases
1. Performance Requirements	1. Performance Requirements
2. Standards of Manufacture	2. Purchase Specification
3. Standards of Quality Control	3. Inspection Specifications

The standards for in-house use can be of the informal variety, but the purchase specifications should be formalized by the joint efforts of engineers and procurement people. The use of such specifications will make their work easier and more enjoyable.

Progress is continuing in establishing standard specifications for the purchase of reinforced-plastic products as well as the establishment of standard methods of tests specific to reinforced-plastic products. The impetus for this progress can be credited to the Society of the Plastics Industry, Inc. (SPI). Credit for accomplishment can be given to many sources, including American Society for Testing and Materials (ASTM), The National Bureau of Standards, scientific and trade associations and organizations, business firms, testing laboratories, and key people in the industry representing producers, distributors, and users. Some of the standards that may be of interest are:

1. NBS Voluntary Product Standard, PS 15–69, Custom Contact-Molder Reinforced Polyester Chemical-Resistant Process Equipment.[1]

This Product Standard covers materials, construction and workmanship, physical properties, and methods of testing reinforced-polyester materials for process equipment and auxiliaries intended for use in aggressive chemical environments, including but not limited to pipe, ducts, and tanks. The Standard is based on the technology of fabrication by hand lay-up or contact pressure molding.

This Standard does not cover, (1) resins other than polyesters, (2) reinforcing materials other than glass fibers, (3) laminate constructions, or (4) filament wound fabrication methods.

2. NBS Voluntary Product Standard, PS 53–72, Glass-Fiber Reinforced Polyester Structural Plastic Panels.[2]

This Voluntary Product Standard covers two types of plastic panels furnished in three weights and seven geometrical configurations, including

1. Superintendent of Documents, United States Government Printing Office, Washington, D.C. 20402 (Catalog No. C13.20/2:15–69).

2. Superintendent of Documents, United States Government Printing Office, Washington, D.C. 20402 (Catalog No. C13.20/2:53–72).

flat panels, intended for use in structural applications. The standard covers requirements for the sizes, configurations, weights, and squareness of the panels. Included are requirements for materials, appearance, color uniformity, light transmissions, transverse load, bearing load, flammability, packing and marking.

The reader will note that the Voluntary Product Standards are specific to polyester resin systems.

Standard Specification for Filament-Wound Reinforced Thermosetting Resin Pipe (ASTM D 2996–71, Pt. 34, 1974 ed.)

ASTM Standards state: "This specification covers machine-made reinforced thermosetting resin pressure pipe manufactured by the filament winding process. Included are a classification system and requirements for materials mechanical properties, dimensions, performance, methods of test, and marking."

Standard Specification for Centrifugally Cast Reinforced Thermosetting Resin Pipe (ASTM D 2997–71, Pt. 34, 1974 ed.)

ASTM Standards state: "This specification covers machine-made reinforced thermosetting resin pressure pipe manufactured by the centrifuged casting process. Included are a classification system and requirements for materials, mechanical properties, dimensions, performance, methods of test and marking.

Standard Specification for Reinforced Plastic Laminates for Self-Supporting Structures for Use in a Chemical Environment (ASTM C 582–68, Pt. 16, 1974 ed.)

ASTM Standards state: "This specification covers the mechanical and physical requirements for reinforced plastic laminates fabricated from a combination of a thermosetting resin and reinforcing materials for use in self-supporting structures where resistance to a chemical environment is the main requirement."

"This specification is not intended to cover selection of the specific resin and reinforcement combinations for specific chemical environments."

Standard Method for Obtaining Hydrostatic Design Basis for Reinforced Thermosetting Resin Pipe and Fittings (ASTM D 2992–71, Pt. 34, 1974 ed.)

ASTM Standards state: "This method describes two alternative procedures for obtaining a hydrostatic design basis for reinforced thermosetting resin pipe, fittings, and piping systems. This method is applicable to reinforced thermosetting resin pipe where the ratio of outside diameter to wall thickness is 10:1 or more and for any practical temperature."

Standard Methods of Test for Interlaminar Shear Strength of Structural Reinforced Plastics at Elevated Temperatures (ASTM D 2733–70, Pt. 36, 1974 ed.)

ASTM Standards state: "These methods cover the determination of interlaminar shear strength of structural reinforced plastics at temperatures above the Standard Laboratory Temperature of 23C (73.4°F.)."

"These methods are not intended specifically for use in determining the effect of long continued exposure at elevated temperatures or of a cyclic exposure range of temperature, although they may be used for such special tests if desired."

Standard Method of Test for Chemical Resistance of Thermosetting Resins Used in Glass Fiber Reinforced Structures (ASTM C 581–68, Pt. 16, 1974 ed.)

ASTM Standards state: "This method is intended as a relatively rapid test to evaluate the chemical resistance of thermosetting resins used in the fabrication of glass fiber reinforced self-supporting structures under anticipated service conditions of use. This method provides for the determination of changes in the properties described in 1 through 4 of the test specimens and test reagent after exposure of the specimens to the reagent:

1. Hardness
2. Appearance of specimen
3. Appearance of immersion media, and
4. Flexural strength.

"The results obtained by this method shall serve as a guide in, but not as the sole basis for, selection of a thermosetting resin used in a glass fiber reinforced plastic self-supporting structure. No attempt has been made to incorporate into the method all of the various factors which may enter into the serviceability of a glass fiber reinforced resin structure when subjected to chemical environment."

The Society of the Plastics Industry, Inc. is continuing its efforts to promulgate additional standards through ASTM Committees. Work is under way on:

1. Fiber glass tanks above ground service
2. Fiber glass tanks for under ground service.

It is suggested that the individual plant or organization use the above mentioned Voluntary Standards, Standard Specifications, and Standard Test Methods as a guide or check list in the preparation of specifications or plant standards. The Voluntary Product Standards can be used as a part of a purchase specification where applicable to the specific type of product. All of these standards serve to promote better understanding between buyers and sellers.

Persons preparing specifications for purchase should emphasize the aspect of performance of the end product and select their specifications on the basis of their contribution to this desired performance.

The information below will provide some assistance to the engineer and/or designer who is responsible for preparation of specifications.

1. **Resins.** The resin to be used should be of a commercial type and should have been evaluated on the basis of demonstrated performance in the specific environment. The purchaser may be better advised to draw on the experience of the manufacturer with regard to the resin selected by specifying the performance requirement. If a gel coat is to be provided, he should spell out the resin to be used, and place reasonable limits on the amount and type of monomer that may be added for viscosity control or for other reasons.

2. **Fillers and Pigments.** The purchaser should set workable limits on the amount of fillers or flow-control agents that can be used, and specify the use of fire retardant agents if they are desired. The use of thixotropic agents in excess of 5 percent by weight will interfere with later visual inspection of the laminate quality. He should specify pigments, dyes, or colorants if desired, but remember that such additions will seriously hamper visual inspection.

3. **Reinforcements.** The purchaser is wise to set forth performance strengths and leave the type and the amount of reinforcement to the manufacturer. If the purchaser is ready to accept responsibility for the strength properties of the laminate then he is entitled to detail the type, weight, and thickness of the reinforcements used to prepare the laminate. The specification should advise the need for a specific surfacing veil, particularly for corrosion-resistant applications.

It may be desirable to limit the maximum build-up of thickness with the addition of a single layer—but on the other hand, the purchaser can exercise control here by specifying minimum shear strengths. The specification should state the minimum thickness for the exposed resin-rich surface and spell out clearly that presence of cracks or crazing will be grounds for rejection. For additional detail, the "Commercial Standard" mentioned earlier in this chapter should be consulted.

4. **Cut Edges.** All cut edges should be coated with resin so that no glass fibers are exposed and all voids are filled. Where greater build-up or thickness is desired on edges exposed to the chemical environment, the coating should be reinforced with chopped-strand glass mat.

5. **Dimensions.** The purchase specification or drawing should state the amount of tolerance that can be accommodated in the laminate section. One word of caution: one should avoid setting exacting dimensions and tolerances just for the sake of making it difficult for the manufacturer. Careful consideration should be given to this detail since reductions of thickness by grinding or sanding are the initial approaches used by the manufacturer to bring his job into specification, and these remedies are not desirable in nonhomogeneous materials. Where dimensions are critical, the custom fabricator is prepared to meet the challenge. Where they are not critical, the manufacturer should be given reasonable latitude. Failure to meet selected dimensions or tolerances should serve as a basis for rejection.

6. **Physical Properties.** The specification should contain minimum values of physical properties to be met. This listing should be a happy medium (long enough to guarantee performance and short enough to keep the price at a satisfactory level). Values for Barcol Hardness related to hours passed after fabrication should always be included. The method of test (ASTM or Federal Spec.) should always be stated along with the physical property. Glass content has also been recognized to be a meaningful test.

7. **Chemical Resistance.** Here is an area where more information is needed. For the consumer, the best approach is to work with a potential supplier and draw on his experience of demonstrated performance, or expose standard-test coupons to the actual environment being considered for the reinforced-plastic application. Drawing on the information accumulated by reputable resin manufacturers can be very much worthwhile from the standpoint of time and money. The purchaser cannot be expected to be a specialist in corrosion resistance. He should expect a recommendation from the supplier as to the appropriateness of reinforced plastics for the specific application. All dealings should be with a supplier who is willing to stand back of his recommendations.

8. **Voids.** The quality and mechanical properties of a reinforced-plastic laminate may be adversely affected by its void content. Some of the suggested effects are lower fatigue resistance, increased susceptibility to water or moisture penetration and weathering, and increased variability in strength properties. The purchaser and the seller should agree on a maximum permissible void content compatible with the end-use requirements of the product.

9. **Testing.** The specification should detail what tests are to be conducted, when they are to be conducted, where and by whom. The purchaser should remember that test sections taken from interior portions (unless there are openings to be cut deliberately) will weaken the structure. The purchaser and fabricator can agree to take samples from the edge of the untrimmed laminate. A clear delineated basis for acceptance or rejection should become a part of the purchase order.

10. **Construction.** Major items can be called out on the drawing or the design submitted by the supplier to the purchaser for his approval. Definition of surfaces should be made. Surface pits greater than 1/8" in diameter and numbering more than 2 per square foot should be cause for rejection. Staggering and lapping of reinforcement layers can be covered in this item. Additional items to be covered will depend on the specific article and accordingly generalization is difficult.

SUMMARY

It is obvious that it is difficult to give detailed advice for a purchase specification to cover the numerous applications (increasing daily) in which reinforced plastics are being used. The inexperienced purchaser would do well to consult with his associates in other companies or with reputable manufacturers. The purchaser should start immediately to accumulate a background of information and, until he has reduced to writing exacting specifications, he may be well advised to direct his efforts to selecting a manufacturer who has a good record in the industry rather than buying on initial price only.

If, on the other hand, the purchaser wants to set up standards and/or specifications (rules of the game), then he should be willing to enforce the specifications through inspection procedures. Specifications that do not relate to needed performance characteristics do nothing but escalate the purchase price. Preliminary discussions with potential suppliers will be enlightening.

Until the individual purchaser or user has accumulated sufficient experience, he may be well advised to limit his specifications as follows:

1. Performance expected.
2. Minimum value for flexural strength.
3. Maximum and minimum glass content.
4. A time schedule of Barcol Hardness values (not ot great value in heat-cured systems).
5. Detailed information on resin-rich surfaces and surfacing veils.

Use of standards of manufacture and performance is definitely a forward step in the reinforced-plastics industry. They reduce the buyers' risk by establishing methods of evaluating reinforced-plastic products. Use of specifications will do much to prevent misapplication and raise the quality of the industry as a whole. Standards should be used to permit the customer to buy in confidence without detracting from progress in the industry.

12

Inspection

Large reinforced-plastic structures, such as covers, ventilating hoods, stacks, and ducts are usually fabricated by hand lay-up techniques. Here the quality of the job and its performance is highly dependent on the workmanship of the individual. There is a need to have a means of evaluating the workmanship, as it affects the service life of the part involved.

Visual inspection provides a simple, easy means of judging the quality of the part. The appearance of reinforced plastic-parts will differ from one manufacturer to the next. Appearance may not always be indicative of serviceability. The type of service for which a part is intended should be carefully considered during an evaluation of visual defects; for instance, a defect in a critical structure such as a tank or process vessel may not necessarily be considered as important in a less critical part such as a ventilating duct.

By placing a strong light behind the wall or section of an unpigmented structure, the inspector can judge the uniformity of the laminate, and detect air bubbles, dry spots, internal cracking, and other conditions which may indicate potential weaknesses in the structure.

The condition of the surfaces may be an indication of the quality of the laminate construction. The inspector should be aware of the application for which the structure is intended; i.e., will both surfaces be in contact or exposed to similar or different corrosive environments? A smooth, true surface is generally a sign of a well-made uniformly cured laminate. If exposed fibers exist on the corrosion surface, there is lack of continuity in the protective resin layer that is so vital to corrosion resistance. Exposed bundles of fiber indicate the likelihood that there is no surfacing mat or surfacing veil. Discoloration may be indicative of inadequate resin mixing or poor-cure practice (exotherms).

Well-prepared industrial specifications often call for cutting a section of the fabricated piece to provide coupons for testing. These cuttings can be taken from overages on length or width or may be produced from the cutting of manholes. The resulting cross section will reveal the depth of the surface layer of resin as well as the shape and conformity of the laminate. Stressed areas often are immediately apparent as internal cracks. The cross section of the test piece will give some indication of the bond between the layers of the reinforcement and the thoroughness of wet-out of the fibers. The location of the surface mat and the thickness of the protective resin-rich layer should be readily apparent.

Another on-the-site test that can be meaningful is Barcol Hardness, but only when used to measure the relative hardness of pieces fabricated from the same system.

If the Barcol Hardness of a laminate falls short of the average ratings found in well-cured laminates of the same resin and reinforcement, then the inspector should immediately suspect an incomplete cure which will later show up in the form of poor corrosion resistance. The Barcol Hardness, to be acceptable to the inspector, should always be greater than the reported hardness (by the resin supplier) of a fully cured, clear (unreinforced) casting.

The inspector should use as a guide the specifications that were part of the purchase order. It is poor policy to include a set of reasonable specifications and then **not** follow through with inspection procedures to see that they have been met.

Summarizing, the inspector should, at a minimum, follow through on the following items:

1. Thorough visual inspection.
2. Uniformity of laminate and adherence to reinforcement specified.
3. Glass content.
4. Barcol Hardness.
5. Specified physical property tests.
6. Overall workmanship.

13

Shop Practice

Quality work can be turned out in a plant maintenance department if there is careful supervision and if the workmen have been trained in a thorough manner. A number of fabrication defects can be avoided or minimized such as voids, wrinkles, delamination, resin dryness or richness, crazing, and foreign inclusion. The presence of any of the above will have an adverse effect on the physical and mechanical properties of a laminate.

VOIDS

Fiber glass laminates often have small voids or air bubbles which detract from their strength properties. These voids are formed generally by entrapment of air during lay-up of the resin and reinforcement and by release of volatile components from the resin system during cure. Voids are never completely absent. An acceptable amount of voids should be established to be compatible with end-use requirements.

Laminates made by contact molding method (hand lay-up) tend to have a higher void content than laminates molded under pressure. Sufficient pressure or careful working of a laminate during its lay-up can force or work out a large portion of the air bubbles.

Air bubbles can be worked out of a lay-up to a considerable degree by use of rollers and other specially designed hand tools. Part of the problem can be minimized by avoiding the inclusion of air bubbles during mixing or stirring of the resin system. Resin cure temperatures should be kept below the point where volatiles from the monomer would form gas bubbles. It is good practice to eliminate the bubbles from one ply before proceeding to the next. Careful placement of plies (working from the middle of the layer to the outer edge, or from one end to the other) will eliminate wrinkles which can result in large voids between the plies. Contacting the edges of an additional ply first and then working toward the center should never be tried—one should work from the center to the outer edge.

116

WRINKLES

Wrinkles in a laminate are caused by the careless handling of the plies of reinforcement during the lay-up or molding process. A wrinkle between the plies of a laminate causes a weak area in the interlaminar bond and reduces the mechanical strength properties of the laminate. A wrinkle in the reinforcement may cause a change of direction of an applied stress which is detrimental to the overall strength of the laminate.

DELAMINATIONS

Lack of intimate contact between adjacent plies in a laminate during cure results in interlaminar separation. The area of separation can be a void space or a pool of excess resin. Either condition results in a weak spot in the laminate. These conditions can be eliminated by proper working of the laminate before proceeding to the next ply. This type of defect does not readily occur in laminates cured under pressure but can readily develop from careless lay-up in the contact method.

RESIN DRYNESS

Laminates are termed *resin dry* when made with insufficient or unequally distributed resin. Resin dryness results in inadequate bond of the fiber glass reinforcement, causing excessive voids or porosity in the laminate. This dryness also invites corrosion attack and contributes to low wet strength retention. To avoid dryness it is better practice to be liberal with the resin.

RESIN RICHNESS

Excessive amounts or uneven distribution of a resin in a laminate can cause resin-rich areas. These resin-rich areas are subject to cracking and will reduce the physical properties in a laminate due to the lack of adequate reinforcement or, in other words, an imbalance of the ratio of resin to the reinforcement. Additional working with tools and rollers will eliminate this problem. It is better to work out the excess resin and perhaps waste a small amount than to leave it in the laminate and degrade the overall quality and workmanship of the job.

CRAZING

Crazing is the formation of tiny hairline cracks through the body of the resin due to rapid or excessively hot curing conditions. Resin-rich areas or heavy unreinforced gel coats of a laminate are subject to crazing. Gradual deterioration in a crazed area of a laminate can be expected to occur when it is subjected to weather and moisture. Laminates made with rigid polyester resins have a greater tendency to craze than laminates made with semirigid resins. In selecting a resin, the purchasing agent should remember that the specification on elongation will define the rigidity of the resin.

FOREIGN INCLUSION

Foreign inclusion results most commonly from careless workmanship. Paper, dust, dirt, pieces of string, and pieces of metal or wood, are all considered foreign to the laminate and reduce its mechanical properties.

Good shop practice requires that a molding operation be kept as clean as possible to prevent contamination of the laminate.

In hand lay-up, it is especially important that each man be his own inspector and judge his work according to the standards of quality set for the job. If workmen are not instilled with this sense of quality control and personal responsibility, the final inspection results in nothing except a lot of rework and additional charges. Lowering standards results in more sloppy work, with no saving in time. A good, smooth, neat job of hand laminating can be done in about the same time as a rough, careless job. Less resin will be wasted and there will be additional time savings in cutting, sanding, fitting, and cleanup.

14

Cure Practices

Because numerous factors influence the cure of polyester resins, a satisfactory cure is obtained only when catalyst and promoter concentrations are properly adjusted for: the overall thickness of the layer to be cured, the gel time required, and the working temperatures.

This section presents a basic guide to adjusting these concentrations for each of the above specialties.

Generally speaking there is a certain amount of flexibility in the choice and concentrations of both catalyst and promoter for use with polyester resin systems supplied by recognized manufacturers. It is also a fact that effective cures are not readily achieved if less than certain minimum concentrations of the ingredients of the curing system are used. These minimums should be detailed by the resin supplier.

ADJUSTING FOR THICKNESS

As polyester resins cure, they develop internal heat, or exotherm. Because of the good insulating properties of a laminate, thicker sections (1/8″ or more) retain much of this heat within the laminate causing it to become warm or hot. Since cure time is related to temperature, the inherent retention of this heat within the laminate will speed the cure of the system. On the other hand, thin sections (under 1/8″) lose much of this heat and do not become warm. As a result, the cure is not accelerated. This heat loss is particularly noticeable when a thin laminate is applied to a metal substrate—a lining to a metal tank, for example.

Because of this heat-thickness relationship, the best cures are accomplished when the concentration of the catalyst and promoter is varied in accordance with the thickness of the part involved.

Thin sections, which do not retain exothermic heat, usually require a greater amount of catalyst to compensate for the absence of temperature increase. For the thick sections (many layers) where exothermic heat is retained, the amount of catalyst can be decreased.

ADJUSTING FOR GEL TIME OR POT LIFE REQUIREMENTS

A range of gel times, or pot life, is attainable for any thickness by varying the amount of catalyst and promoter. It is important to remember, however, not to go below the recommended minimums.

ADJUSTING FOR TEMPERATURE OF RESIN AND WORKING AREA

The gel time of a resin system (with any given concentration of catalyst and promoter) will vary with the temperature of the work room and the temperature of the resin. Most gel times are reported at 70°F or 75°F. An ambient working temperature of 60°F should be the minimum permitted if good cures are to be achieved.

CURING SHRINKAGE

During the curing process, the thermosetting polyester resins used in reinforced plastics shrink in volume due to molecular cross-linking when passing from the liquid state to the solid state. This reduction in volume or polymerization shrinkage is more marked in polyester resins than in epoxy resins. During the curing process the resin forms a bond with the reinforcing glass. Since the glass does not undergo any appreciable change in volume during the cure, shrinkage of the resin sets up compressive stresses in the glass and residual tensile stresses in the resin. In addition to these stresses (which are acting at cross-purposes) the interfacial bond between the glass and resin develops a shear stress during the curing process.

Another type of internal stress occurring during cure is due to thermal shrinkage. The curing process as mentioned previously develops a considerable amount of heat in a laminate due to the exothermic chemical reaction. Resins have a considerably higher coefficient of thermal expansion than glass. A differential thermal shrinkage between the glass and the resin occurs when a laminate cools after curing. The effect of thermal shrinkage is additive to that of polymerization shrinkage in causing residual stresses in the glass and resin of the cured laminate.

The residual stresses in the complex structure of a fiber glass laminate can affect its short-term or long-term loading strength, fatigue strength, and resistance to crazing and weathering.

Since these stresses and potential defects result from the curing process, they can be minimized by selection of optimum curing methods. Excessively rapid cures or exotherms higher than those required for proper cure should be avoided since they can cause damage to a laminate during cure. Thick laminate sections are particularly subject to uneven cure and thermal stresses.

The above emphasizes the importance of arriving at the optimum level of catalyst and promoter to accomplish the desired results. One principle that could be applied to most situations would be to try to select as long a curing time as is economically feasible.

The assistance of the resin and catalyst manufacturers should be solicited in establishing the optimum curing cycle for the molding process until the fabricator has established methods for his own shop based on extensive shop experience.

15

Cure Systems

Below is a table showing percentages of catalyst and promoter for room temperature cure (75°F) of a typical polyester resin. For this particular resin system, the manufacturer recommends that the concentration of cobalt naphthenate should not be less than 0.25 percent and that the concentration of methyl ethyl ketone peroxide (MEKPO) should not be less than 0.50 percent. Note that the gel times or the usable pot life can be controlled by varying the percentages of MEKPO and cobalt naphthenate promoter.

Laminate Thickness	% MEKPO 60% Active	% Cobalt Naphthenate 6% Active	Pot Life Minutes	Time to Stripping
Up to 1/8″	2.0	1.0	15–20	40–50
	1.0	1.0	20–30	60–80
1/8 to 1/2″	1.0	1.0	20–30	60–80
	1.0	0.75	50–70	90–110
Over 1/2″	1.0	0.5	70–110	Several Hours

The above information is typical of a manufacturer's recommendations. Variations to the curing cycle are subject to many variables, such as:

1. Ambient temperature of working area.
2. Humidity.
3. Type of resin.
4. Thickness of laminate.
5. Age of resin (possible gel time drift with promoted resins).
6. Promoter system (age will decrease activity).
7. Inhibiting and/or accelerating effect of accessory items such as pigments, colorants, and fillers.

In the case of the first item, a rise of $10°F$ may reduce the pot life significantly. Also a variation in the temperature of the polyester resin of 10 to $15°F$ away from $75°F$ will significantly affect the cure times.

In very cold weather the polyester resin should be allowed to warm up slowly to the working room temperature. This period of time should be a minimum of 24 hours.

The effect of temperature on the curing of a resin system is extremely important and a good deal of attention has been focused on it by resin manufacturers and fabricators. If the work room or fabricating room is not adequately heated in cold weather, the catalyst requirements may vary significantly from summer to winter. Often a fabricator will take advantage of a temperature disparity between outdoor temperature and that of indoors. A cooler area is better suited for lay-up operations where a longer pot life is desirable, and a warmer location is better for speeding up the curing cycle. Direct sunlight is a further aid to rapid curing. The use of heat lamps, heated rooms, or other aids has found widespread acceptance.

A small oven will lessen the curing time of fabricated units by heating moderately.

Postcuring in a heated room or oven is desirable, but no single rule is generally applicable. For small items, a postcure of one hour at $200-225°F$ will enhance the properties of the laminate. Postcuring is particularly desirable where the laminate is to be used in corrosive environments.

Using a Barcol Hardness tester is a convenient means of checking on the progress of the cure of a system. Barcol Hardness readings of 30 to 60 or better are attainable but will vary with resins from different manufacturers.

Some laminates made with well-known corrosion-resistant resins will exhibit lower Barcol Hardness than others made with conventional polyester resins. A Barcol Hardness of 30 to 40 after curing 24 hours at $75°F$ would be typical. After one week the Barcol Hardness will level off at 40 to 45.

Thick laminates or those containing thixotropic agents will show a slightly higher Barcol reading (3 to 5 units). Resins systems containing paraffin, synthetic fiber overlays, or separately cured gel coats may show readings of 3 to 5 units less than the standard laminate.

A practical way to follow the cure is to take Barcol Hardness readings every 24 hours after the laminate has reached a tack-free stage of cure and plot their values against time. The readings will level off in 144 to 168 hours.

PREPARATION OF PROMOTER AND CATALYST SOLUTIONS

Promoter

Cobalt naphthenate (6 percent solution as purchased commercially) can be used without dilution when the quantities needed are large enough to permit accurate measurement. To insure accuracy when adding small amounts, a 10 percent stock solution in styrene (10 parts by weight in 90 parts by weight of styrene) can be made up for use as needed. This solution should be expected to be stable for approximately one month.

Disposable plastic syringes available from local drug stores are most convenient for measuring quantities of 5 cubic centimeters or less.

Catalyst

Methyl ethyl ketone peroxide (MEKPO) is the most common catalyst used for hand lay-up fabrications. It is available as a liquid and is normally purchased as a 60 percent active material. The diluent is 40 percent dimethyl phthalate.

WARNING—Catalyst must never be mixed with promoter or promoter with catalyst. Violent decomposition may occur.

Mixing Catalyst and Promoter Solutions into the Resin

As cautioned above, catalyst and promoter must always be added separately to the polyester resin and one must be thoroughly mixed in the resin before the other is added. Separation of ingredients and a restricted mixing procedure are necessary because of the combustible character of the mixture.

A propeller-type mechanical agitator is necessary to properly disperse the promoter and the catalyst in the resin and to prevent "hot spots" in the finished laminate. Each ingredient must be mixed for 3 to 5 minutes to insure homogeneity.

It is desirable to add the promoter to the resin system, and the catalyst later.

TACK-FREE CURES

Polyester resins are available with or without wax. These resins would be described as nonair-inhibited and air-inhibited, respectively. An air-inhibited polyester will exhibit a sticky or tacky surface when cured in air.

A nonair-inhibited polyester (containing wax) offers the user a tack-free cure without utilizing special techniques.

An air-inhibited polyester is recommended for use in hand lay-up for all layers except the top layer. The absence of wax permits additional layers to be applied and provides for a better bond between adjoining layers. If by chance the fabricator has used a polyester containing wax and he desires to build up additional layers, he must let the system cure, and then remove the wax from the top layer by sanding with 120-grit sandpaper.

The fabricator can use air-inhibited resin systems and attain a tack-free surface after cure by placing a layer of cellophane (cellulose acetate 5 mils in thickness) on top of the last layer to prevent air contact with the resin surface. This technique takes a little additional time but pays off in the form of a very smooth top surface.

Generally speaking, the small fabricating operation would be best advised to purchase air-inhibited resin and use the film covering technique or add a wax material as described below.

The addition of a 0.4 percent by weight of paraffin to the resin system provides nonair inhibition and a tack-free cure. The paraffin is generally added as a styrene solution. Household paraffin wax, available at local food stores, is suitable for this use. It is best used by preparing a (stock) 10 percent solution of wax in warm styrene (10 parts of wax by weight in 90 parts of styrene). A quantity of 4 percent of the 10 percent stock solution at room temperature added to the resin system will provide the desired tack-free cure. The wax solution should be added to the resin before the promoter and the catalyst.

It should be stressed again that the use of paraffin wax has an adverse affect on adhesion and it should not be used until the final layer of the resin system is added to the laminate.

ADDING FLOW-CONTROL AGENTS

Polyesters applied to vertical surfaces tend to flow or sag before they have had time to gel. This condition can be remedied by the addition of 2 to 3 percent of colloidal silica. Cab-O-sil, a product of the Cabot Corporation, is a well-known and accepted product which will provide thixotropic properties to a resin system.

Cab-O-sil is a light, fluffy material which requires considerable mixing for complete dispersion. It should be added to the resin gradually and with constant agitation. The agitator blades should be brought up near the resin surface to increase the mixing shear at that point. When properly wet out, Cab-o-sil will not separate from the resin mix on standing.

Another colloidal silica that has been accorded good acceptance is "SANTOCEL" 54, produced by Monsanto Chemical Company.

Colloidal silicas must be kept dry in storage if they are to be satisfactory for use.

Should the fabricator desire to use styrene for viscosity control, he should always assure himself that the styrene is free of polymer. Polymerized material will not produce a satisfactory resin system. To test for polymer content, take a test tube containing approximately one milliliter of styrene monomer and add approximately 10 milliliters of methyl alcohol. If the solution remains clear, it is polymer free and satisfactory for use. If it is cloudy on mild shaking, the styrene contains polymer.

16

Laminate Thickness

The material in this chapter forms a basis for estimating the thickness of a laminate made by hand lay-up methods. Using the examples presented one can estimate the raw material requirements and costs.

Thicknesses and weights per unit area of conventional reinforcing materials (dry basis) are listed in suppliers' catalogs.

In laminating operations, the addition of the polyester resin to a reinforcement (fiber glass) will increase the thickness by a factor of about 15 percent over the thickness of the dry material. Exceptions to this rule are the case of the surfacing veil, where the resin-rich layer gives a significant increase in thickness to the fiber glass material; and the case of the chopped-strand mat, where the thickness may decrease in relationship to the amount of compacting. The amount of resin that a particular reinforcement will hold will vary with its porosity, its position in the laminate, and the characteristics of the resin being used.

To fabricate the flat panels below by hand lay-up methods estimate material requirements by the following method:

Ten panels 48″ × 96″ with a finished thickness of about ¼″ with this reinforcement construction:

Layer Number	Description	oz/ft^2 Glass	Thickness Mils	Average Glass Content %
1	Surfacing veil	0.1	10	15
2	Reinforcing mat			
	Continuous strand	1.0	17	30
3	Reinforcing mat			
	Continuous strand	1.5	25	30
4	Woven roving (5 × 2.5)	3.0	62	45
5	Reinforcing mat (Chopped)	1.0	26	30
6	Woven roving	3.0	62	45
7	Reinforcing mat (Chopped)	1.5	38	30
8	Cloth 16 x 14	1.1	14	50
		$12.2 \ oz/ft^2$	254 mils	34.4% average

Note: Thicknesses of individual layers are approximate.

Based on the above example, the laminate will contain 12.2 ounces of glass per square foot with an average glass content of 34.4 percent by weight. On this basis, the weight of resin will be 65.6 percent or 23.2 ounces per square foot, and the laminate will have a total weight of 35.4 ounces per square foot.

The designer may want to check his calculations. Fiber glass has a specific gravity of 2.54 or weighs 0.0917 pounds per cubic inch.

A typical resin has a specific gravity of 1.13 and weighs 0.0408 pounds per cubic inch.

At a ratio of 34.4 percent glass to 65.6 percent resin, the laminate will weigh:

$0.0917 \times .344 = 0.03154$
$0.0408 \times .656 = \underline{0.02676}$

.05830 lbs. per cubic inch

or

0.9328 ounces per cubic inch

144 square inches 254 mils thick would have a volume of 36.58 cubic inches. At 0.9328 ounces per $inch^3$ the calculated weight of the proposed laminate is 34.1 ounces. This is well within the limits of accuracy of the method presented.

Flat sheets made by pressure-molding techniques show a somewhat higher glass content and weigh in the neighborhood of 1.1 pounds per square foot for 125 mils of thickness. A panel of 254 mils would have a calculated weight of 36 ounces per square foot. Use of fillers can vary the weights considerably.

The estimator should determine the amount of resin required to complete the lay-up of the 10 panels. At 23.2 ounces per square foot and 320 square feet total the amount of resin would be $\frac{23.2 \times 320}{16} = 464$ pounds. Because of waste, the estimator should add a minimum of 15 percent to the base requirement $(1.15 \times 464 = 534$ pounds). At 9.4 pounds per gallon the amount would be 57 gallons.

In estimating the cost of the raw materials, the widths of reinforcing materials that are available must be considered and the wastage of glass kept to a minimum. In addition, the reinforcing should be ordered so the laminate can be built oversize, 50″ x 98″ at a minimum, and later trimmed to the proper size.

In the example used, the reinforcement is available in 50″ widths. Using the prices posted by suppliers, the designer can develop the dollar cost of glass, remembering that the glass must be cut and that the final laminate must be trimmed. In the case of the resin system, 15 percent has already been included for waste. The estimator, after developing the cost for polyester resin, should add an additional 15 percent to cover costs for catalysts, mold release, and solvents. Special gel coats or pigments will add to the final cost of the raw materials.

In all probability the plant requiring relatively small amounts of flat sheets will learn that it is better to purchase their requirements of flat sheets from recognized reputable suppliers rather than to hand fabricate. For requirements involving small-area laminates, irregular in size or uneconomical to cut from flat sheets, or of unusual thickness, one may find that hand fabrication in the plant shop is the better approach. The information that has been presented should help guide this decision. The labor cost has not been included in this example because of the wide variances that can occur due to the level of experience and the subsequent factor of productivity. A minimum amount for labor for this job would be equal to the raw material cost.

In designing a laminated structure, the engineer or designer is concerned with strength properties. Generally speaking, parts built by hand lay-up methods do not develop strength properties to the same effectiveness as reinforced plastics fabricated by other techniques.

To give the designer a point of departure, Lahde and Moore[1] presented information to answer the question "How much resin should be worked out of a boat hull lay-up?" It is true that when the glass content of laminates is increased, the unit strength (psi) is also increased. Laminates were prepared in four different thicknesses and compared. The reinforcement in each laminate consisted of 6 oz. cloth, 1½ oz. mat, and 18 oz. woven roving.

Panel No.	Laminate Thickness	Glass Content %	Flexural Strength psi	"Simulated Breaking Force" (lbs)
19	0.157	22.8	35,500	291
4	0.133	25.8	38,800	226
18	0.110	27.0	40,700	163
26	0.072	41.4	56,000	97

Since the term *simulated breaking force* may not be clearly understood, the authors are quoted verbatim: "Laminate thickness was decreased from 0.158″ to 0.072″. In general the unit flexural strength increased gradually as the amount of resin was removed to make the panel thinner. However, the actual breaking force and not the unit strength is the limiting factor. For example, figuratively speaking, panel 26 with its 56,000 psi flexural strength and thickness of 0.072″ would be broken by a '97 pound log'; whereas, Panel 19 with a strength of only 35,500 psi and a thickness of 0.157″ would require a 291 pound blow. In other words, decreasing a laminate thickness decreases

1. SPI–RP Div., *Proceedings*, Vol. 15, Sec. 7–D, 1960.

its 'utilitarian strength' in spite of increasing its strength per square inch [of cross section area] ."

Lahde and Moore also studied the effect of the type of reinforcement with respect to its position in the laminate. On the test work they used various reinforcements; i.e., cloth, woven roving, and mats ranging from 1 ounce per square yard up to 3 ounces per square yard.

The results of their work confirmed that strength properties can be varied by (a) selection of the type of reinforcement, and (b) the location of a particular type of reinforcement in a multilayered laminate.

More detailed information can be found in the original paper printed in the proceedings of the 15th Annual Technical Conference of the Society of the Plastics Industry, Reinforced Plastics Division, Section 7–D.

SPI reports the following information on representative properties of hand lay-ups:

	Hand Lay-Up	
Property	*Mat Reinforcement vs.*	*Fabric*
Sp. G.	1.4 to 1.7	1.6 to 2.0
Glass content %	30 to 40	45 to 55
Tensile strength, psi	10,000–20,000	30,000–50,000
Tensile modulus, psi x 10^6	0.8 to 1.8	1.5 to 4.5
Compressive strength, psi	15,000–25,000	30,000–56,000
Flexural strength, psi	20,000–40,000	45,000–75,000
Flexural modulus, psi x 10^6	1.2 to 1.8	2.0 to 4.0
Impact strength, ft-lbs per inch (notched)	5–25	20–30

There is considerable variation in properties in either type of reinforcement and, therefore, reproducibility of strength properties in hand lay-up operation may be a problem.

Johns-Manville reports the following comparison of properties of laminates prepared with woven rovings at different molding pressure. The reported strengths of the warp and fill directions are averaged below:

Molding Pressure	*Contact*	*12 psi*	*30 psi*
% Glass content	57.1	64.4	70.2
Tensile strength, psi	39,100	39,100	43,000
Flexural strength, psi	80,250	75,450	66,850

Oleesky and Mohr, in their recent book *Handbook Of Reinforced Plastics,* page 486, list some typical hand lay-up constructions:

No.	Laminate Construction	Thickness Mils	Glass Content %	Flexural Strength psi	Modulus of Elasticity x 10⁻⁶ psi
1	1½ oz. mat 8 oz. cloth	70	30.0	33,200	1.40
2	2 oz. mat 24 oz. W.R.[a]	125	27.5	46,500	1.07
3	6 oz. cloth 1 oz. mat 24 oz. W.R.	125	27.4	44,300	1.51
4	6 oz. cloth 1½ oz. mat 25 oz. W.R.	125	30.2	45,500	1.51
5	10 oz. cloth 2-2 oz. mat 10 oz. cloth	149	22.6	27,600	1.24
6	10 oz. cloth 2-24 oz. W.R. 10 oz. cloth	104	43.6	23,700	1.66
7	¾ oz. mat 24 oz. W.R. 3 oz. mat 18 oz. W.R.	125	39.6	55,900	1.68

a. W.R. stands for "woven roving."

Strength determinations were made by loading on top surface; i.e., No. 1 on the 1½ oz. mat surface.

The above examples would indicate that thickness, composition, and glass content provide the designer with the basis for improved flexural strength, and this is absolutely true in a test panel. The designer should remember, however, that in the fabrication of a structural item, such as an enclosure, a more economic means of obtaining stiffness is to include some reinforcing members (structural shapes) in the direction where increased flexural strength is required.

17

Making a Simple Hand Lay-Up Laminate

The discussion below may seem elementary to the average reader, but it may serve a useful purpose in training maintenance personnel if the trainee actually performs the operation of making a laminate 10″ x 10″ from 6 layers of glass.

Materials required:
8 square feet of cellophane (24″ x 48″) (cellulose acetate, 5 mils)
1 square foot of surfacing mat
2 square feet of reinforcing mat (chopped strand)
1 square foot of glass cloth 16 x 14 weave
1 square foot of woven roving
1 quart of resin
1 ounce MEKPO (60%)
10 ml. of paraffin solution
Necessary handtools, mixing cups, stirring sticks, roller, squeegee, brush, protective equipment, measuring devices, masking tape, etc.
Solvent for cleanup—acetone, xylol or methyl isobutyl ketone

For purposes of demonstration and instilling confidence into the worker, it is assumed that the working room temperature is 75°F and that the resin contains a promoter in the amount of 1 percent. The worker should plan to add the volume of catalyst that will permit 15 to 20 minutes working time.

1. Cut the different types of glass into pieces 12″ x 12″. Cut the acetate film into 2 pieces 24″ x 24″.

2. Using a smooth table top or a piece of plywood, fasten down the cellophane to the plywood base with the masking tape. Take care to remove all wrinkles.

3. Mix sufficient resin and hardener to do the job. Until experience is

developed, the supervisor can use a rule of thumb of about one gallon of resin per 40 square feet of laminate. It is suggested the worker mix one pint of resin with hardener. This will be an excess amount but it is impractical to mix less than a pint accurately.

4. After the resin and catalyst has been thoroughly mixed, apply a generous layer to the film with the brush. Cover an area at least 12″ x 12″.

5. Carefully lay down the first layer of surfacing mat or veil. Using the roller and additional resin, work the resin into the fiber glass surfacing mat. The operator will note a visual disappearance of the fabric pattern of the glass mat as it becomes wet out.

6. Lay down the 12″ x 12″ piece of reinforcing mat and apply additional resin, using the roller to work the resin into the mat until it is thoroughly wet out. The operator will observe that additional working as well as additional resin is required to wet out the mat as compared to the surfacing material that was used as the first layer.

7. Continue the operation by placing a second layer of mat, adding resin, and working same into glass reinforcement.

8. After adding additional resin, place in the laminate a layer of cloth fabric (fiber glass) and repeat the working and rolling operation to remove air bubbles and wet out the cloth.

9. Next add a layer of woven roving, adding more resin to wet out the reinforcement. The operator can continue to add additional layers to increase thickness if desired.

10. If it is decided to stop at this thickness, then add a final layer of resin. Follow this by carefully placing on the laminate the remaining piece of cellophane. Smooth the surface gently with carefully applied pressure working out all air bubbles.

11. Pick up the piece of plywood and place laminate in the sun or in a room where the temperature is greater than 75°F.

12. Clean the tools that were used. Observe the amount of resin used. (It may be necessary to catalyze additional resin to complete the job the first time.) Personnel should clean any resin from hands, preferably by washing with soap, water, and cleansing powder. A rag dampened with acetone or xylol may be necessary to remove resin from hands.

The following sketch illustrates the various layers and the sequence in which they are applied, starting with the bottom layer.

In the above example, the resin system used was air-inhibited so it was necessary to cover the top of the layer to keep the air from inhibiting or preventing the cure.

The worker should have gained some feel for:

1. Mixing the resin and catalyst.
2. The amount of resin required.

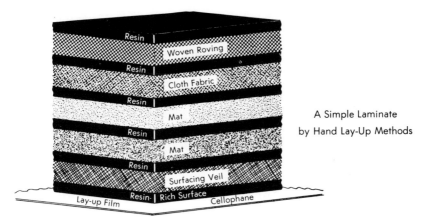

A Simple Laminate
by Hand Lay-Up Methods

Thickness of layers exaggerated for illustrational purposes.

3. The difference in ease of wetting of each of the three types of glass fabric.
4. The removal of air bubbles from each layer as well as the air bubbles when the layer of film was placed over the final layer.
5. The tendency of a layer of the laminate to travel when pressure is applied in a single direction.
6. The importance of prompt clean-up.

Assuming there will be several workers to be trained, the instructor can further the knowledge of the class by having them observe the progress of the cure of the laminates that have been made. It is suggested he uncover the first laminate that was prepared in about two hours after completion and inspect additional laminates at 2 to 4 hour intervals. If a Barcol Hardness tester is available, the progress of the cure should be measured by taking readings over a period of 168 hours (every 12 to 24 hours after the first 8 hours).

Should the instructor care to demonstrate the use of paraffin to illustrate tack-free cures, he can modify the above example by having one laminate covered to keep air from contacting the resin surface, one left uncovered to let the air contact and thus present a tacky surface, and finally add an appropriate amount of paraffin solution to the final resin layer of one of the laminates and let it cure uncovered.

Laminate with Gel Coat
Consider the making of a simple laminate by hand lay-up methods where a gel coat is involved. To keep the example simple, the sheet of acetate film is assumed to be the flat mold surface.

The first layer of resin laid against the mold surface is known as the gel coat. It may have one or more of the following functions: provide a decorative layer, provide an extremely smooth surface, provide improved impact resistance, or provide corrosion resistance as well as weathering resistance.

Some fabricators elect to formulate their own gel coats while those not well experienced choose to purchase the material already formulated. Depending on the desired thickness of the gel coat, the fabricator may elect to provide some amount of reinforcement by the use of a surfacing veil. An appropriately pigmented gel coat permits the use of unpigmented resin for the remainder of the laminate. The gel coat should always contain a thickening agent and should be catalyzed to have a short gel time. If it is to be applied by spraying, then the viscosity must be adjusted appropriately. For those not equipped with a spray gun, the gel coat can be applied by brush, but this technique is usually less than satisfactory. Apply using criss-cross strokes to provide a uniform film of 15 to 25 mils. Just before gelation occurs, the surfacing mat can be laid on the surface of the gel coat to provide reinforcement. Once the gel coat has hardened, then the fabricator can proceed to add layers of resin and reinforcement as described in the earlier example.

Cloth or mat can be used to reinforce the first lamination of a lay-up system. Woven roving is not generally used because its coarse weave pattern may be reproduced (photographed) in the cured gel coat. The inexperienced molder is well advised to use an open-weave cloth. It is relatively easy to evacuate the air from these open-weave fabrics. They drape well and flatten readily with hand pressure. When the open-weave cloth is applied at the proper interval, the tackiness of the previous coat or layer makes the cloth adhere snugly to the surface of that previous layer.

Where a pigmented gel coat has been used (white or pastel) and a surfacing mat is to be laid against it, it may be desirable to give the gel coat a very thin coat of quick-setting black pigmented resin immediately after the gel coat has cured to a stiff gel. Air bubbles (voids) entrapped in the laminate will show up clearly against this background. This quick-setting black pigmented coat can be used to provide the tackiness needed for the layer of cloth mentioned previously. When chopped-strand mat is used directly behind the gel coat, special care must be taken to eliminate entrapped air. When mat is laid into a layer of wet resin, it gets a partial wetting-out from the underside. Additional lay-up resin is then applied to the top of the mat with a wool-covered roller or dabbed on with vertical strokes of a paint brush. Resin is best applied from the top of the piece down, and from the middle of the area out to the edges.

There are several methods by which mat is compacted in hand lay-up operations to evacuate air from the laminate. When mat is alternated with cloth in so-called sandwich construction, the cloth is used as a screen through which the air is eliminated without displacement of the mat fibers. The air in the saturated mat is forced forward by the scrubbing action of a rubber squeegee against the cloth. Mat can also be compacted with rollers. These can be of the conventional type or of grooved construction, such as would be obtained by alternating large and small diameter discs of metal or fiber on an axle with a suitable handle.

Whenever possible, a laminate should be completed in a continuous operation to insure the best physical and mechanical bond between laminations. If it is necessary to stop the laminating, the fabricator should complete the wetting-out of the last layer of reinforcement before terminating his work. On returning to the laminating process, the operator should make sure the surface is free of fiber protrusions before proceeding. Any fibers projecting out from the surface can be removed by light sanding. Light sanding over the entire surface tends to improve the bond between the laminates.

In making a flat sheet lay-up, the reinforcement projecting beyond the desired finished size (area) can be trimmed with a sharp knife as soon as the laminate has attained a stiff enough gel to be cut. Delay in trimming will require later use of a saw to cut the laminate to size. Any sanding operation should be delayed until the laminate is thoroughly cured. The fabricator should remember that the laminate is not dimensionally stable until the cure is relatively complete and any handling should be done with care. Use of external heat, moderately applied, will of course expedite the curing action.

18

Mold Making

Some readers may want to construct a simple master mold (tooling) which will permit the production of a number of molded parts. Agreement on the procedure for making a production mold using glass-reinforced polyester will never be unanimous. In this chapter some of the fundamentals concerning the master pattern, gel coating, laminating, mold thickness, materials, and a step-by-step procedure are discussed. The information presented is intended as a guide and can be modified as required to suit the individual fabricator's special needs or circumstances.

MASTER PATTERN

Wood is the most practical material for constructing the master pattern. Making the master pattern can best be left to industrial pattern or model-makers. An actual part existing in fiber glass-reinforced plastic or any other material, providing its surface is impervious to styrene, the monomer that is present in polyester resins, can serve as the master pattern for making the mold. The selection or making of the master pattern must be done with care. The part later to be produced from the new mold will not only produce the desirable features of the master pattern, but also any undesirable qualities, surface defects, or flaws.

For purposes of this discussion, it is assumed that the master pattern is completely acceptable to the fabricator. At the risk of being elementary, it is also assumed that undercuts are absent from the shape and geometry of the master pattern and that the pattern has sufficient draft so it can be removed from the master mold without damage to the mold or to the pattern.

If the master pattern is constructed of wood the surface should be sealed with shellac or a plastic coating. Some prefer to seal the wood pattern with a polyester-resin coating. In situations where the pattern may be used to produce a number of production molds, it is common practice to use glass-surfacing veil with polyester resin and actually build up the surface of the pattern by a thickness of 30 to 40 mils. Let us assume that the wood pattern has been sealed with shellac. Over the sealant, apply three coats of lacquer primer, sanding after the application of each coat. After sanding the third coat, apply a final coat of satin lacquer. Final sanding should be done using a progression of 320- to 600-grit paper. Wet sanding will give a better surface than dry sanding. Follow sanding by hand buffing for maximum smoothness.

After buffing to desired smoothness, the surface must be waxed—usually with Carnauba wax. Three coats of wax should be applied with buffing following each wax application to be sure there is no wax build-up.

Some molders prefer to spray the waxed master pattern with a release agent such as polyvinyl alcohol (PVA) or with silicone. These release agents are likely to reduce the gloss of the mold and of the subsequent finished production part.

After waxing and final polishing the pattern is ready for the application of a polyester gel coat. The application of the gel coat is the first step in the actual construction of the mold or tooling.

GEL COAT

The gel coat is basically a thin film of highly pigmented and filled polyester resin. It gives a smooth, glossy, durable surface to the mold. Production gel coats serve as decorative surfaces which require minimal maintenance while offering protection against ultraviolet degradation of the structural laminate. A mold or tooling gel coat offering high heat distortion is always desirable. The gel coat also serves to mask the structural details or patterns of the reinforcement. Since the proper cure of the gel coat is important, follow carefully the directions of the gel coat manufacturer with respect to catalyst concentration. A catalyst concentration of 2 percent MEKPO is not uncommon. A concentration in excess of 3 percent to 4 percent may be detrimental to proper cure. If a future layer of substrate resin is applied before the gel coat is fully cured, it will attack the relatively thin gel coat layer and result in a wrinkled or "alligatored" surface. Overcure of the gel coat will reduce the adhesion between the gel coat layer and the subsequent layer of glass-reinforced structural back-up.

Selection of the mold or tooling gel coat should be made with considerable care. Tooling gel coats differ from production gel coats basically in hardness and rigidity (mold gel coats are more rigid than production gel coats). Some fabricators prefer to use a black pigmented gel coat; others, two layers (a clear layer plus a pigmented layer). Where molds are going to be used many times in production, it is good practice to use a black gel coat for the first layer and an orange gel coat for the second layer. This pigmentation system readily indicates the amount of wear that the mold has been subjected to during refinishing or repair. Another advantage of the two-layer pigmented system is the increased visibility of air bubbles or entrapped air. Entrapped air must be kept to a minimum.

The gel coat should be spray-applied to the master pattern to approximately 30 to 40 mils of thickness in two applications of 15 to 20 mils each. The first layer should be allowed to cure for approximately one hour prior to the application of the second layer. Application of the gel coat by brushing is seldom satisfactory. A gel coat thickness of 30 to 40 mils will allow the necessary depth for any sanding or polishing of the mold after the master pattern is removed.

Some of the companies that market gel coats are: Cook Paint and Varnish Company, Ferro Corporation, Glidden Durkee Div., SCM Corp., Koppers Company, Inc., and RAM Chemicals Div., Whittaker Corp.

LAMINATING RESIN

The polyester resin selected for completing the lay-up of the mold should have a low viscosity to facilitate fast wetting of the glass reinforcement while permitting the release of any entrapped air.

To insure minimum shrinkage and exotherm during cure, the fabricator should adjust the catalyst concentration to provide gel times of 30 to 45 minutes. An exception to this gel time is in the application of the skin-coat layer. The instructions of the resin manufacturer should be followed carefully when determining the concentration of catalyst to be used.

There are many companies which market polyester laminating resin. A partial list of producers of polyester resins is listed in the Appendix. Technical information on products may be obtained on request from the manufacturer.

For the purpose of this discussion, one laminating resin formulation that can be suggested is:

Koppers Polyester Resin 6060-5	100 parts by weight
MEKPO	0.4 to 0.5 parts by weight

For the skin-coat layer (after finishing the gel coat) use a catalyst concentration of 1.0 percent.

REINFORCEMENT AND THICKNESS

Since the amount of reinforcement to be used is dependent on the thickness of the mold to be constructed, it is worthwhile for the beginner to determine in advance the approximate desired thickness. The average thickness may be in the range of $3/8''$ to $1/2''$. The total thickness as well as the number of layers required to achieve that build-up is affected by the size and geometry of the master pattern. When long, unsupported distances are to be spanned, the total thickness must be greater to obtain dimensional stability and rigidity. Once the mold is placed in production, it must prevent distortion due to the weight of the part being produced as well as distortion caused by the exothermic reaction during the cure of the production parts.

Braces or stringers are necessary on large molds to provide the desired rigidity and dimensional stability. Angle iron, steel tubing, or glass-reinforced polyester structural shapes or tubing can be used. Wood members for bracing should be considered with care since changes in humidity may cause dimensional change in the wood. The bracing may be incorporated with the last few layers of build-up by providing adequate gusseting at the point of contact with the mold.

Before the lay-up is begun, it is good practice to cut or tailor the reinforcement to the size and shape needed. This affords better economic utilization of the reinforcement. Small pieces left over from the tailoring can be used for added build-up at critical areas or for gusseting of the bracing.

Selection of the specific reinforcement to be used may be varied within reasonable limits depending on the size of the mold. Generally speaking, the layers of reinforcement farthest from the gel coat surface can be of greater bulk and thickness. The use of woven roving to achieve a greater laminate thickness per layer applied usually results in the fiber pattern of the roving printing through to the gel coat surface. If this occurs the fiber pattern will also be reproduced in every subsequent production part.

For the simple mold, the practical choice is 1½ oz. glass mat plus 7-oz. or 10-oz. glass cloth fabric. The layer of reinforcement next to the gel coat can be 7-oz. or 10-oz. cloth fabric. Some fabricators prefer to use a layer of surfacing veil (reinforcement) next to the gel coat. If surfacing veil is used, it should be followed with a layer of glass cloth.

The following example illustrates how to estimate the number of layers of reinforcement to give the indicated thickness. The example is intended as a guide and may be varied to suit the specific situation.

Layer	Approximate mils, thickness
Gel Coat	30–40
10-oz. cloth (skin layer)	16–20
1½-oz. mat	38–50
10-oz. cloth	16–20
3 layers of mat and cloth	162–210
	262–340 mils

During the placement of the reinforcement into the laminating resin, it is extremely important to eliminate air entrapment and keep resin-rich areas to a minimum. Resin-rich areas increase linear shrinkage and stress the gel coat which may cause irregularity in the finished mold surface.

MATERIALS REQUIRED

Mold or tooling gel coat with recommended catalyst and manufacturer's
   use instructions
Laminating resin (promoted) with recommended catalyst and manufacturer's use instructions
1½-oz. glass mat (chopped strand or continuous strand)
10-oz. glass cloth
Braces or stringers
Spray gun, necessary hand tools, mixing containers, measuring devices,
   stirring paddles, brushes, rollers, squeegees, etc.
Protective equipment
Solvent for clean-up

For estimating purposes, the amount (pounds) of laminating resin required
will be about 3 to 4 times the weight of glass used.

PROCEDURE

1. Apply the catalyzed gel coat to the waxed master pattern by spraying,
building up to a thickness of approximately 30 to 40 mils. Application should
be done in two layers of 15 to 20 mils each with at least an hour of waiting
between layers.

2. Allow the final layer of gel coat to cure till it is barely tacky but does not
release gel to the finger touch. One test method is to take a conventional ball point
pen and determine if you can make a mark on the gel coat with it. If the gel coat
is adequately cured, it will accept writing from a ball point pen. If the gel coat is
not cured, the pen will not leave an ink mark. Careful attention must be given
to accomplishing the desired cure of the gel coat. If not sufficiently cured, the
styrene monomer present in the laminating resin (next layer) will attack the gel-
coat film. The gel coat is air-inhibited (containing no wax) and the surface will
remain slightly tacky even when fully cured. This tackiness serves to improve
the bond between the gel coat and the subsequent layer of laminating resin. The
tackiness also serves to locate positively and hold in position the first layer
(skin coat) of reinforcement prior to proceeding with the laminating.

3. Carefully lay the skin-coat reinforcement of 7-oz. (or 10-oz.) glass cloth
onto the cured gel coat. The resin for this skin-coat layer should have a catalyst
concentration of 1.0 percent MEKPO. This concentration of catalyst will reduce
the gel time to approximately 15 minutes. Brush catalyzed laminating resin onto
the cloth, working the resin through the reinforcement and thoroughly wetting
all fibers (note the disappearance of the fabric pattern) and working out all
entrapped air. The roller or squeegee can be used to advantage at this point. It
is imperative that the first layer of reinforcement make total intimate contact

with the gel coat. The skin-coat cloth layer must be mitered at corners or complex curves where wrinkles can occur. Watch out for bridging of the fibers in corners or in indentations that may be designed into the pattern. Continue to check, up to the time the resin gels, for air bubbles or entrapped air. If air voids are observed, they should be cut out with a sharp trimming knife. Extreme care must be used when cutting the cloth so there is no damage to the gel coat. If small areas of reinforcement are removed, cut pieces of cloth to replace them. Laminate these pieces into the cut-out areas using laminating resin. Unless these areas of replacement are carefully completed, the end results will be unsatisfactory. A great number of patches make for a subgrade mold with short life expectancy. It is better practice to give the necessary attention to eliminating all air bubbles as the skin coat is applied, rather than spending even more time patching and repairing. Once the skin-coat layer has cured, reinspect the layed-up area for voids and repair as indicated.

4. Once the skin-coat layer has cured (approximately 4 hours at 70°F.), you are ready to continue the lay-up using laminating resin (catalyzed to give a 30 to 45 minute gel time) with the application of 1½-oz. mat, plus 10-oz. cloth for each layer. Here again it is important to work the reinforcement into the resin and to push out all entrapped air and to maintain a uniform layer of laminating resin to avoid resin-rich areas. Cure of this layer should take about 4 hours. A Barcol hardness (impressor) may be used to establish progress of the cure. Readings of 30 to 40 would indicate sufficient surface cure. The ultimate Barcol hardness for a good tooling resin is in the range of 45 to 50.

5. Repeat step 4 (after allowing approximately 4 hours for curing), adding another layer of laminating resin plus the reinforcement of mat and cloth.

6. Repeat step 4 until the desired thickness of the mold is obtained. Prior to placement of the final layers, it is good practice to install the bracing and stringers using resin and reinforcement to gusset them to the mold. Cardboard formed as a "V" section can be laminated into the exterior surface of small molds to improve structural rigidity. If stringers are laminated to the exterior surface of the mold, it is good practice to place a cushion of cardboard between the stringer and the surface to prevent reproduction of the stringer (print through) in the gel-coat surface of the mold and subsequently to the part that will be made in the new mold.

7. After the final layer is laminated, allow the mold to cure at room temperature for at least 48 hours. Oven-curing (after 24 hours of room temperature curing) at no more than 150°F. for 12 to 24 hours will accelerate the cure. Thick sections (greater than 1/4 inch to 3/8″) should be cured for longer periods. The master pattern should be left in the mold during the curing period. The master mold should not be used for production until it has cured at least 168 hours.

8. After the mold has been adequately cured, the master pattern can be removed. Care should be exercised to avoid damage to the pattern as well as to

the master mold. Some fabricators build air ports or water ports into the master pattern or into the mold to assist in part removal. Building ports into the mold is considered the better practice. The number of ports is influenced by the size of the mold. The use of 1/8″ steel tubing (flare-type fittings) is suggested.

9. After the master pattern has been removed, flashing and excess material can be removed from the mold with portable hand tools. Final removal of material to the trim line should be accomplished by machine or hand sanding.

10. Inspect the surface of the completed mold. Small flaws or defects can be repaired with gel-coat resin or polyester patching putty. On defects less than 1/8″ diameter, use catalyzed gel coat after sanding the surface of the defect to assure good adhesion of the gel coat. For flaws over 1/8″ diameter use tooling resin mixed with milled glass fibers (1 part catalyzed resin to 3 parts of milled fibers) to fill the flaw. After the putty has cured the area must be covered with gel coat to provide the desired surface gloss.

11. If the surface of the mold does not have the desired cosmetic surface, wet sand or use automotive type rubbing compounds and polish to desired smoothness.

12. Before using the new mold for production, be sure to wax it. Apply two layers of wax after each use for the first five uses. Once the mold is broken in, a light application of wax after each use suffices. If the finished surface of the molded part is not critical, a release agent can be used as a substitute for the wax or in addition to the wax coating.

Should the reader desire more detailed information on making molds, a number of excellent publications are available from the suppliers of reinforcement materials.

Mold for large one-piece thirty-five foot boat hull. (Owens-Corning Fiberglas Corporation)

The procedure for building a mold outlined above is probably more suited to the needs of the beginner rather than to the needs of the production shop. It is common practice in industry to use spray-up techniques to build up the entire thickness of the mold rather than using hand lay-up as has been described.

19

Making a Contoured Laminate

Some readers may require a contoured laminate. A typical application would be the fabrication of a housing or enclosure to serve a particular need or a housing for the repair of mechanical equipment (a blower or fan).

If the dimensions and geometry of the completed laminate are critical and if a number of production parts are required, then the best approach is to use a master pattern and construct a production mold following the principles outlined in the previous chapter.

Whether the mold is female or male depends on which of the two exposed surface areas is to have the better finish and appearance. If the better surface finish is to be on the interior of the contour, then a male mold is desirable; but if the better surface finish is to be on the exterior, then the female mold is the appropriate choice. Where corrosion-resistance considerations are involved, it is strongly recommended that the smooth resin-rich surface be the surface area that is to be exposed to the corrosive environment. Another consideration may be the removal of the molded laminate. Generally speaking, the completed laminate can be removed more easily from a female mold than from a male mold because of the normal shrinkage of the polyester in the completed laminate.

If the need is for one contoured laminate, or a limited number, and if there is reasonable freedom with respect to dimensions and geometry, then an inexpensive mold may provide the desired end results.

An inexpensive mold can be constructed by building a support, covering the support with chicken wire, mesh, or metal lath, and applying a gypsum plaster. The wet plaster can be formed manually to give the desired contour or profile. Normally one inch of plaster build-up is desirable when designing the support form. After cure, the plaster can be sanded to the final desired shape. The plaster mold must be sealed once the plaster is completely dry. Shellac or plastic lacquer can be used. The principles applying to the release of the laminate from the mold as previously discussed must be followed. If the surface finish of the final laminate is not critical then the surface of the mold may be covered with cellulose acetate film (cellophane) and held in place with acetate film, pressure-sensitive tape.

In maintenance and repair work with reinforced plastics a laminate having a shape or geometry that will not lend itself to removal from a conventional mold may sometimes be needed. In this situation the choice is either to subdivide the assembly into several parts that can be removed from the mold or to construct an inexpensive mold and destroy or break it away from the finished laminate.

The reader desiring a more detailed discussion on making tooling or molds from gypsum plaster should refer to the *Handbook of Reinforced Plastics,* Oleesky and Mohr, Chapter V–3, pp. 328–355.

One of the desirable applications for reinforced plastics is the repair and maintenance of housings, cylindrical objects, or other mechanical devices that have lost their material continuity, but are still structurally satisfactory. Repairing tubes or cylindrical devices can be done in a manner similar to the method described in chapter 20, "Field Repair of Pipe."

The following approach to the repair of a fan or blower housing that has lost material continuity but is structurally satisfactory may be of interest to the fabricator since contoured surfaces are generally involved.

It is desirable to clean the metal housing thoroughly by sand blasting. Once this is completed, expanded metal lath can be shaped and fitted into the housing, and tack-welded to hold it securely in place to form a base for the application of a hand lay-up laminate. Some have found it desirable to use chopped-strand mat cut in circles one inch in diameter and staggered on 4- to 5-inch centers. The circles of mat can be bonded to the underside of the metal lath with polyester resins. Once this has cured, the metal lath is welded through the open areas to the metal substrate with the glass mat sandwiched between the metal lath and the metal housing undergoing repair. Once the lath is in place, the fabricator can proceed to make a laminate over the metal lath using a thickened resin system and additional layers of glass mat. The fabricator can select the different types of reinforcement to give the final desired surface.

An alternate approach used with success is to weld the expanded metal to the housing prior to the start of the hand lay-up. Once the lath is in position, the fabricator can place an oversized piece of mat or woven roving over the lath and, using a blunt stick of wood (dowel), force the reinforcement into the openings of the lath. It is probably sufficient to force the reinforcement into the lath in a predetermined pattern, filling about 10 percent of the holes. Once the reinforcement is in position, the fabricator can proceed with the completion of the laminate by hand lay-up techniques.

Be sure to be generous with the resin on the initial layer of reinforcement so that the fibers projecting into the lath will be thoroughly wetted. The fabricator will observe that the completion of the laminate in this particular case is built up in reverse order with respect to the other examples that have been presented. This does not create any insurmountable problems, but it does make additional working of the laminate surface necessary to achieve the desired smoothness of the final layer.

In corrosion-resistant applications the final layer can be a resin-rich layer of polyester that in many respects is similar to a gel coat as previously described.

20

Field Repair of Pipe

In the reinforced-plastics industry, the terms *pipe* and *tubing* are often used interchangeably. For purposes of this discussion the following definitions are suggested to differentiate between the two:

Pipe. Generally speaking, pipe is used in the transport of fluid materials under pressure, whereas tubing is used as a structural support, such as a column. The working pressures of pipe are specified by the manufacturer. For small diameters, design pressures of 300 psi are common, while diameters exceeding 3 or 4 inches up to 10 inches may be rated at 150 psi. Diameters from 12 inches to 24 inches may have pressure ratings at 100 psi. The foregoing does not disqualify the use of pipe for transporting fluids at low pressures; i.e., drain lines.

Tubing. Tubing (generally made by processes other than filament winding) can be used to transport fluid materials at relatively low pressures, say 50 psi or less. Wall thickness is generally greater than for pipe of similar diameter. Tubing may be used for mechanical purposes such as structural supports.

For purposes of this discussion, field repair of pipe is recommended only where the pipe has been damaged by impact or abrasion but has not lost its continuity. Where the pipe has been severed, the section should be replaced with new pipe and appropriate fittings, unless it is an application in which pressures of less than 50 psi are involved.

Satisfactory field repairs can be made by the methods described for pipe at its rated pressures, up to 4″ in diameter. For larger diameters additional strength will have to be built into the repair if the system is to be returned to service at its rated working pressures.

The following materials will be useful in performing repairs:

Adhesive resin
Mixing cups, stirring sticks, and brushes
Cleaning solvent and paper towels

Roll of fiber glass cloth tape
Laminating resin
Cellophane
Masking tape
Sandpaper
Hacksaw
Shears or razor blade

1. Using sandpaper, abrade the area to at least 2 inches beyond the repair point, continuing around the entire circumference of the pipe. In the case of epoxy pipe, clean the area with solvent-saturated paper towels. In the case of polyester pipe, wipe away the dust created by the sanding action. If the area becomes contaminated (by grease or oil, etc.) after sanding, then clean the area that has been sanded with solvent.

2. Prepare bonding adhesive and laminating resins. Follow manufacturer's directions carefully. Prepare only as much material as can be used within the stated "pot life" of the resin system.

3. Liberally apply a uniform depth layer of the bonding adhesive over the entire cleaned area of the pipe. Be sure the pipe is properly supported with at least 4 inches of clearance above ground.

4. Apply one layer of fiber glass tape by rolling around pipe in a manner similar to applying a bandage. Work the tape into the bonding adhesive using a brush. Make sure of close, even contact between the tape and adhesive.

5. Now using the laminating resin system, continue to apply the tape by winding in the same direction, and, using the brush (same brush as used with adhesive), apply liberal quantities of the resin. It is important to "wet" the fiber glass tape (each layer) completely before proceeding with the next layer of tape. By working the resin into it, the fiber glass tape becomes "translucent." The number of layers to be applied is suggested below:

Pipe Size	1"	1½"	2"	3"	4"
Layers	7	7	7	9	9

After applying the desired number of layers (each layer thoroughly wet out), cut the tape with the shears or razor blade. If the area to be repaired is wider than the width of the fiber glass tape, it will be necessary to wrap the tape spirally to insure sufficient coverage for the repair. Make sure that all areas have at least the minimum number of layers as shown above. Be sure that the last layer is thoroughly wet out.

6. Wind three layers of cellophane tape on top of previously applied fiber glass tape. Wind in the same direction as used in Steps 4 and 5. Apply firm pressure, or wrap tight, while winding with cellophane tape to smooth the surface and work out any entrapped air. Cut off cellophane tape and secure in position under tension, with masking tape.

7. Allow full time to harden or cure according to the manufacturer's recommendation. In the case of epoxy, the time may vary from 2 to 8 hours depending on the temperature. The time for polyester may be somewhat shorter depending on the amount of catalyst used. This curing time may be shortened by applying heat (moderately) from an external source. After cure is completed, the cellophane tape may be removed.

8. Return the line to service and inspect the repaired section.

21

Patching with Reinforced Plastics

In this chapter, the techniques that can be used to patch or mend a surface area that has become discontinuous because of corrosive attack, mechanical damage, or a combination of both are briefly outlined. The techniques may be used (in scaled-up proportions) to patch large areas.

The supervisor is cautioned to check his design to be sure that adequate strength has been provided if the part to be repaired is under pressure, or vacuum, or is subject to tensile or compressive loading stresses (weight-bearing). Suggestions are based on the use of polyester resins, but epoxy resins may be used equally well in this type of patching technique.

1. Remove as much of the damaged area (exposed or visible side) as practicable in order to provide sound material all around the damaged portion.

2. Use a power sander (disc type) on the inside (nonvisible) to remove all surface finish or deteriorated material over an area 4 to 6 inches back from the perimeter. Continue sanding until sound material has been reached. A rough surface finish is desirable. Try to featheredge the material at the perimeter of the area to be patched.

3. Using a clean rag dampened with xylol, acetone, or methyl ethyl ketone solvent, clean both surfaces (inside and out) of the sanded area to remove any oil, dirt, or other contamination. Remember to use proper precautions when working with flammable solvents.

4. Prepare a temporary backup for the patch. This may be masonite, plywood, stiff cardboard, or sheet metal faced with polyethylene film, cellophane, waxpaper, or a suitable release agent. Tape the patch over the hole and provide structural support to retain the backup plate in place.

5. Prepare the required amount of fiber glass reinforcing material, cutting the layers 6 inches larger than the hole to be patched to provide sufficient overlap. Cloth fabric or woven roving is preferred for patches for comparatively flat or gently curved surfaces. Nonwoven mat is generally preferred when the

area to be patched involves complex curves or contours. Remember that mat is lower in strength than cloth—but that it is more economical. The availability of mat in a number of different thicknesses, however, may make it more desirable to use. Here, one should consider in advance the number of layers and the composition of the laminate he is planning to use as the patch.

6. If the area to be patched is comparatively small, the operator may decide to prepare a so-called preformed laminate by laying up the entire laminate on cellulose acetate film. See the discussion in chapter 17, *"Making a Simple Hand Lay-up Laminate."* In using this technique, one should thoroughly wet out each layer of the laminate and squeegee or roll to remove all air bubbles. When the laminate patch is the desired thickness, pick up the laminate, using the cellulose acetate film as a support, and place the laminate onto the area to be patched. Work the laminate to the desired contour, making sure that the laminate conforms to all depressions and making sure that all air bubbles have been worked out.

If the area to be patched is too large to be repaired by this method, it may be advisable to build up the laminate to the desired thickness using the base item as the support along with the back-up plate. In this case, coat the surrounding area heavily with activated resin and use fiber glass cloth as the first layer. Work the cloth tight against the base material, using sufficient resin to thoroughly wet out the cloth, and removing all air bubbles. Cutting the layer of reinforcement 8 to 10 inches larger than the hole to be patched, but leaving the outermost 2 to 4 inches dry, can work to advantage. By using this approach, the operator can use cellophane tape or masking tape to help hold the reinforcement in place. Tape from the dry reinforcement to the sound material.

Continue to build up the laminate to the desired thickness by adding additional layers of reinforcement and resin. If a polyester resin—air-inhibited—has been used, it will be necessary to place a film of cellophane over the last layer to obtain a tack-free surface.

7. After the patch has hardened, strip off the film and remove the back-up material. Time of cure will depend on the temperature and the amount of catalyst used with the polyester resin. Auxiliary heat, such as infrared lamps, may be used to speed the cure. Auxiliary heat, if used, must be carefully applied. Use of heat prior to gelation of the resin will cause it to become temporarily more fluid which will result in runoff (vertical surfaces) and in some cases actual shifting or slipping of the laminate patch.

8. If the operator decides to make the repair in two or more steps he must make certain that he obtains a satisfactory bond between the first layer and subsequent layers. If a nonair-inhibited resin (one containing wax) has been used or if a glossy surface is present, then this wax or glossy surface must be removed. Sand to a depth to remove the gloss but not to a depth to raise the glass fibers.

9. If the final appearance of the patch is of prime importance, then the

final surface of the laminate can be sanded, using sandpapers of increasingly fine grit followed by wet sanding with 200- to 400-grit paper.

10. If the inner surface of the patch is to be exposed to a corrosive environment and the repaired item is of a size sufficient to permit a workman to enter and work, then it may be desirable to prepare the inner (nonvisible) surface and add a surfacing mat as a resin-rich interliner. In such a case the layer of mat should overlap the base material.

22

Fabrication Practices

Fabrication is the making of a finished article (structure) from sheets, tubes, rods, or other reinforced-plastic (RP) shapes. The methods and techniques currently being used for working with reinforced-plastic shapes may be of interest.

Fabrication with RP can be compared in many ways to working with wood—with some differences in cutting and fastening.

Ordinary hand or power tools can be used in most cases. When fabricating large structures or large quantities, special tools such as carbide or diamond-tipped saw blades are recommended for longer tool life and faster cutting speeds.

GENERAL SUGGESTIONS

1. Common safety precautions should be observed. For example, the operator of a circular power saw should wear a face mask to protect his eyes.

2. A coverall or shop coat will add to the operator's comfort during the sawing, machining, or sanding operations. The dust created can cause skin irritation; the amount of irritation will vary from person to person. It can be reduced or eliminated by use of a protective cream. To avoid inhalation of the dust, an appropriate face mask should be worn.

3. Machine ways and other friction-producing areas should be cleaned frequently. The combination of grease and fiber glass chips can become a damaging abrasive if allowed to accumulate.

4. Excessive pressure should be avoided when sawing, drilling, routing, etc. Too much force can rapidly dull the tool.

5. Excessive heat must not be generated in any machining operation because it softens the bonding resin in the laminate which results in a ragged rather than a clean-cut edge.

6. The reinforced-plastic material should be supported rigidly during cutting operations. Shifting may cause chipping at the cut edges.

7. The use and design of fastening devices for mechanical connections should be carefully considered.

8. For adhesive fastening *the surface should be prepared properly for bonding* prior to the application of the adhesive.

9. The strongest connections of high reliability can be made by using a combination of mechanical fasteners with adhesives.

10. Any cut surfaces or edges of the reinforced-plastic shape should always be touched up or sealed before the job is reported complete.

Following are a few suggestions regarding fabrication that may prove helpful:

MACHINING

Sawing or Cutting

Always provide adequate support to keep the material from shifting when making a cut. Without adequate support reinforced-plastic shapes or profiles will chip.

In cutting operations, use light, evenly applied pressure. Heavy pressure tends to clog the teeth of the blade with dust particles, and the cutting life of the blade is thus shortened.

Water cooling is desirable when many pieces or when thick cross sections are being sawed. With cooling, cutting speeds increase, smoother cuts result, and dust is largely eliminated.

Straightline Sawing. Deciding to saw RP with a hand or power saw depends, among other things, on the quantity being cut. When sawing a relatively few pieces, a disposable-blade hacksaw (24 to 32 teeth per inch) is suitable.

Straightline sawing can be accomplished quickly and accurately with a circular power saw. A table or radial model is better than a portable hand model because of the built-in rigidity and guides which insure accurate cuts. However, a hand model can be effective too—especially if the blade is set as deep as possible.

For infrequent cutting on a circular power saw, a metal blade with coarse, offset teeth can be used satisfactorily. For frequent cutting, a masonry saw blade—preferably carbide-tipped—will give accurate cuts and reasonably long blade life. For production cutting, use a 60- to 80-grit diamond-tipped blade for best results.

Circular power saw equipped with a diamond-coated blade for production cutting RP structurals.

One problem that may be encountered with a circular power saw is that of not being able to cut large sections in one pass due to the blade size. However, many large sections can be sawed in two passes by cutting halfway through from one side, inverting the material, and cutting the other side.

If the cross section is too large for the circular saw, two-pass method, or if large sections are being sawed in quantity, a power bandsaw with a carbide or diamond-tipped blade should be used—preferably a machine with automatic feed to insure a light, even pressure on the blade. In the cutting of tubing a smoother cut can be made if the tube is "rolled" through.

Circular or Curvature Sawing. Good results can be obtained by using a saber saw or band saw on small quantity cutting. Neither machine is particularly recommended for production sawing unless carbide or diamond-tipped blades are used to avoid excessive blade replacement.

A band saw is ideal for cutting circles and irregular shapes.

A hand router with rotary bit can also be used to cut circles and curves, but it removes considerably more stock.

In cutting rod or bar stock, a hacksaw may be convenient to use. A blade with 24 to 32 teeth per inch is effective for hand cutting; light rapid strokes should be used.

Abrasive blades (carbide or diamond), which may have become clogged because of overheating or too much pressure, may be cleaned by cutting a common brick.

Shearing

Shearing is not recommended unless the shear is equipped with a specially shaped blade that allows only a small portion of the cutting edge to penetrate the material at any one time. Even then, shearing will not be as precise as sawing, but it is a fast cutting method. *Do not* shear RP shapes over ½" thick.

Punching

The best possible punching results can be obtained by using a specially shaped punch that permits only a small part of the cutting edge to penetrate the material at any one time. The punch should be equipped with strippers so it can be removed without cracking the material around the hole.

As a general guide, punched holes run .002" to .005" smaller than the punching die, while punched blanks run .001" to .005" larger. Die clearance should be about half of that used for steelworking dies. *Do not* punch fiber glass-reinforced laminates over ½" thick.

Drilling

Any standard twist drill is an excellent tool for working fiber glass laminates. Carbide-tipped drills *are not* necessary, even when cutting large quantities, because an ordinary high-speed steel drill will give sufficiently long tool life. Drill speeds should be roughly equivalent to those used for drilling hardwood. When drilling large holes, a backup plate of wood will prevent the hole from breaking out on the back side.

Important Note for Close Tolerance Work: Holes drilled in RP are generally .002″ to .004″ undersize. Thus, a 1/8″ drill *will not* produce a hole large enough to admit a 1/8″ expanding rivet. Instead, a No. 30 drill must be used.

For very large holes, a hole saw is preferable to a fly cutter.

Routing

Both hand-held and bench-type routers give excellent results. Rotary file bits—preferably carbide tipped—are best when a large amount of routing is being done. Two-fluted wood bits can also be used, but they require frequent sharpening and are therefore practical only for occasional routing.

Caution: Use light pressure when making a cut. Forcing the routing operation causes the resin to heat up and soften—and the bit may be damaged if it becomes clogged.

The stationary router is particularly useful for cutting keyways and performing other milling operations. Note vacuum hose for removing dust and small fiber glass chips.

Turning

Most metalworking machine tools can be used in working with fiber glass laminates. Tool steel cutters—both single or multiple point—are entirely satisfactory for short-run machining operations on small quantities. Carbide tool bits, such as Carbaloy 999 or Wiley's E–3, are recommended whenever a great deal of machining is to be done.

In general, dimensional tolerances should match cold rolled steel tolerances; feeds and speeds should match those used for brass or aluminum.

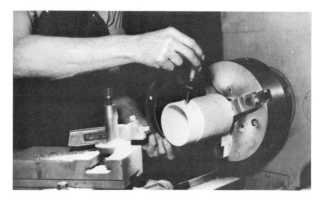

When turning RP on a lathe, set the cutter slightly above center to reduce tearing action on the glass fibers.

The best machined finish can be obtained by climb-cutting instead of undercutting. Reason: climb-cutting reduces the tearing action on the glass fibers. A water coolant will aid in giving a good machined finish, too.

Round nose lathe tools also provide good finishes. The tool should have very little clearance. A single point tool tends to tear the material and also results in round corners rather than sharp corners. Surface speeds should be adjusted to give the desired finish and are determined by the hardness of the material and the type of cutting tool in use. The work should be fed continuously and steadily, for if the tool is stopped in the middle of a pass, the material will be noticeably marked.

Threading and Tapping

Threading of RP is not recommended as a means of mechanical fastening where high strength is required. It should be avoided in the design of fabricated components whenever possible. This is because the threading operation cuts the continuity of the glass fibers and leaves only the shear strength of the resin component to provide the strength of the thread.

Threaded connections are satisfactory where strength is not an important consideration.

Bonding of the threaded connection with a polyester or epoxy adhesive will improve the strength of the connection.

Standard taps and dies can be used for threading. Plain or soapy water should be used to lubricate the cutting. If the joint is to be bonded later with adhesives, then only plain water should be used.

When tapping a blind hole, be sure to allow adequate clearance at the bottom of the hole to prevent the tap from bottoming and damaging the threads.

Grinding

Grinding is generally not recommended. Centerless grinding of tubes and rod can be done satisfactorily if specialized equipment is available. In ordinary grinding operations, the dust tends to load the stone and stop the grinding action. If grinding is required, use an open grit wheel and water as a coolant.

Sanding

Open-grit sandpaper on a high-speed sanding wheel gives best results. Use very light pressure—*do not* force the sander against the fiber glass surface because heavy pressure may heat up and soften the resin. Wet sandpaper applied by hand or with an orbital sander will produce a high gloss finish.

MECHANICAL FASTENING

Mechanical fasteners, when carefully selected and properly applied, can be used to advantage in making connections between two pieces of reinforced plastic or between reinforced plastic and some other material of construction. Many available standard fasteners are suitable if the necessary modifications to joint design are made prior to their use.

In developing the design of the joint, due consideration must be given to the following:

1. Glass-reinforced materials generally have lower shear and bearing strengths than most metals.
2. The joint must be designed to prevent damage to the plastic surface when the joint is under load as well as during installation.

For a complete discussion of the use of mechanical fasteners, other texts should be referred to, such as *Handbook of Reinforced Plastics,* by Oleesky and Mohr, Chapter VII-1, page 381.

These two authors have proposed the following criteria for the design of bolted or riveted joints:

1. In designing a mechanical joint, the number of fasteners, fastener size, and spacing should be selected so that the joint is critical in bearing. This means that as the load on the joint is increased, the bearing load of the laminate is exceeded before reaching the tensile strength of the reinforced plastic, or the shear strength of the fiber glass reinforced material, or the fastener. By doing this, catastrophic failure of the joint in tension or shear is prevented.

2. Stress concentrations exert a dominant influence on the magnitude of the allowable design tensile stresses in reinforced plastic joints. Generally speaking, only 40 to 50 percent of the ultimate tensile strength of the plastic laminate is retained in a mechanical joint.

3. Multiple rows of fasteners are desirable in reinforced-plastic joints. These rows permit a more gradual load transfer between the joined members and reduce stress concentration effects. Multiple rows of fasteners are required in unsymmetrical joints (lap joints and single shear butt joints) to counteract the bending induced by eccentric loading.

4. The local reinforcing of a joint to increase its tensile strength should be avoided because only small increases in load carrying capacity are obtained and the increased eccentricity in unsymmetrical joints gives rise to greater bending stresses.

5. Since stress concentration and eccentricity effects cannot be calculated with a consistent degree of accuracy, it is advisable to verify all critical joint design by fabricating and testing representative sample joints.

Definitions used by Oleesky and Mohr to describe specific joint failures are: "Tensile failures in mechanical joints in reinforced plastic parts start with formation of a small crack at the edge of the fastener hole. The crack then spreads rapidly to the part edge or to an adjacent hole. Shear failures occur by pulling out a material slug between the fastener hole and the end of the laminate. or by pulling the fastener head through the laminate. Bearing failures (deformation to 4 percent of the hole diameter) are not catastrophic, but fracture finally occurs by tension if the stress persists. Simultaneous shear tear-out and bearing failure can occur in joints with multiple rows of fasteners."

Nailed Connections

Nailing is a satisfactory way of fastening fiber glass-reinforced laminates to wood and to other materials that provide enough grip to hold the nail. Common nails can be driven through 1/16"-thick RP without predrilling holes; tempered nails will go through 5/16"-thick material. RP heavier than 5/16" requires predrilled holes, slightly oversize, to admit the nail and to allow for expansion and contraction between the reinforced plastic and the material to which it is nailed. It is also advisable to predrill slightly oversized holes before nailing long lengths of thinner sections of laminates. **Never nail RP to RP.**

Screw Connections

Self-tapping screws have been used successfully in many applications involving mechanical connection where high-strength fastenings are not required. A better use of self-tapping screws is in combination with adhesives. In this application the screws can serve to hold the adhesive bonded surfaces of the two parts together while the adhesive cures. The screws also contribute limited mechanical strength to the connection. Appropriately sized pilot holes should be provided for the screws. In corrosive environments, stainless steel or Monel screws should be used unless a suitable coating of polyester or epoxy can be applied to the exposed screw heads to prevent discoloration of the area due to rusting.

Lag screws are not recommended because they do not take a good bite in reinforced plastics. If lag screws must be used, then a predrilled tapered hole must be provided.

Bolted Connections

Very satisfactory connections can be made between reinforced plastic components by using standard machine bolts, nuts, and washers. Since reinforced plastics can fail under high localized stress conditions, such as those encountered around a bolt, the tighter the bolt is in the hole—using appropriate washers to place the member under compression and to distribute the load—the more effective will be the connection.

The strongest joint or connection between pieces of reinforced plastics is obtained by using a combination of properly fitted bolts with adhesives applied to the appropriately prepared mating surfaces. Corrosion-resistant bolts should be used where environments so indicate.

Bolting into Tapped Holes

Mechanical fastening can be done by screwing bolts into tapped holes. As stated earlier, the mechanical properties of tapped holes are not good and the connection will not be particularly strong. Sheet metal screws can be used for removable cover plates. Strength of the connection can be improved by the use of threaded inserts bonded into place with suitable adhesives.

When removable bolts are required, threaded metal inserts or fasteners should be installed in the reinforced plastic, preferably bonded into place with a suitable adhesive. Threads in reinforced plastics will wear out quickly and may not provide sufficient holding strength. Many types of metal inserts and fasteners are commercially available, for example: "Molly" nuts, "Tee" nuts, "Dzus" fasteners, B. F. Goodrich "Rivnuts," and "Helicoils." Some types need to be bonded in place; others can be mechanically fastened.

Another way of installing removable bolts is by tapping the reinforced plastic, applying epoxy or polyester adhesive in the hole, and inserting the bolt after covering the threads and shank with grease or some other releasing agent. The bolt can be withdrawn after the adhesive has formed and hardened around the threads. This method is *not* recommended when an exceptionally strong connection is required.

When bolts are to be installed permanently, a tight connection is easily made by tapping the reinforced plastic and applying epoxy or polyester adhesive to the hole just before inserting the bolt. Be sure that the bolt is not contaminated with grease or oil which would serve as a release agent.

Riveted Connections

United Shoe Machinery Corporation "Pop" rivets are very effective in joining RP sections. These rivets are available in various sizes and heat styles in aluminum, steel, Monel, copper, and stainless steel. Other types of rivets, such as drive rivets, formed by a rivet gun, or the conventional rivet, formed with a ball peen hammer, can produce an effective mechanical connection. The strength of the connection can also be improved with suitable adhesives. Backup washers are recommended for distributing load stresses. As in drilling opera-

tions, it is necessary to use a slightly larger drill than the exact diameter of the rivet. For a 1/8" rivet, use a No. 30 drill rather than a 1/8" drill.

Helpful Hints, a handbook on fastening with screws, nuts, and bolts, available on request from Russell, Burdsall and Ward Bolt and Nut Company, is of considerable interest.

ADHESIVE FASTENING

Adhesives can provide strong and durable bonds between reinforced plastics or between RP and other structural materials. Satisfactory bonds are obtained if the joint is designed to avoid excessive peeling stresses, if the mating surfaces are properly prepared, and if the recommended types of adhesives are used.

The two types of adhesives recommended for use with fiber glass reinforced materials are polyesters and epoxies. Either adhesive will produce a satisfactory joint; however, polyester adhesives are somewhat less convenient to use because of the difficulty of measuring the small amount of catalyst required.

The "pot life" of different adhesives will vary with the quantity mixed at one time and with the temperature. Small quantities mixed at room temperature will be usable over a period of one hour or more, while quantities of one pint to one quart may harden in one-half hour or less. It is recommended that only the amount of adhesive that can be used up during the pot life of the material be mixed. Hardening time or "pot life" can be extended by placing the mixed adhesive in a refrigerator until ready for use.

Surface Preparation

Before reinforced plastic shapes can be bonded or glued, *the surface must be properly prepared* to insure proper adhesion.

Contaminated surfaces should be thoroughly cleansed by wiping with a clean rag dampened with a solvent such as acetone, toluol, or methyl alcohol. Do not immerse or soak RP shapes in these solvents.

Making the Adhesive Joint

1. Remove surface film left during manufacture by sanding both mating surfaces using 120-grit sandpaper. Sanding is adequate when surface gloss has been removed. Avoid sanding to a depth that will raise the glass fibers. On large surfaces, coarser grit sandpaper may be more practicable to use.

2. Remove any dust remaining on surface to be bonded from sanding operation by using an air blast or wiping with a clean dry rag. Avoid recontamination of the surface by handling.

3. Mix adhesives according to recommendations of manufacturer.

4. Spread a thin film of adhesive on both surfaces that are to be joined, making sure to cover any glass fibers that have been exposed.

5. Appropriately clamp assembly to permit adhesive to harden.

Remove surface film. Sand both mating surfaces to
be bonded.

Remove dust from sanding by air blast or by wiping with
clean rag.

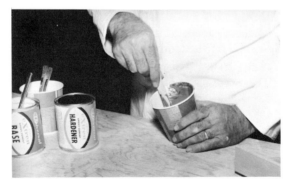

Follow manufacturer's directions carefully in preparing
adhesive.

Spread thin coat of adhesive on both surfaces to be joined.

Clamping to hold bonded pieces in place while adhesive hardens.

Curing the Adhesive Joint Using Clamps to Maintain Bond Pressure

Freshly bonded joints should be held in position with clamps or weights until the adhesive cures. Joints bonded with epoxy adhesives generally can be handled with reasonable care after 8 hours. It is desirable to leave the clamps or maintain the bonding pressure on the joints overnight or for a total of 20 to 24 hours. If an oven is available, the curing time can be lessened considerably by heating moderately. The structure should not be expected to carry its design load until the adhesive joints have cured a minimum of 48 hours at 70°F. Lower temperatures will require longer cure times.

Using Mechanical Fasteners to Maintain Bond Pressure

Self-tapping screws, bolts, or rivets can also be used to hold freshly bonded joints in place. Mechanical fasteners provide these important advantages: (1) no waiting for the adhesive to cure—one can proceed at once with further fabrication—and (2) increased stress reliability at the joint.

Hold bonded joints with mechanical fasteners to permit
handling assembly before adhesive hardens.

Cleanup

The adhesive should be cleaned from hands and tools before it hardens.
Solvents such as xylol, acetone, or methyl ethyl ketone are suitable. Gasoline is
not effective. Remember to use proper precautions when using flammable
solvents.

Taping of Joints

In the fabrication of some items, butt joints may be present. The overall
strength of the joint or connection, as well as the appearance, may be improved
by taping with fiber glass tape or strips of fiber glass mat together with polyester
or epoxy resin.

In the case of making tape joints, the surface must be properly prepared
by the methods recommended for adhesive joints.

Taping

PAINTING AND FINISHING

Most reinforced plastics are resistant to weathering and chemicals and, in general, do not require paint for protection. It may be desirable to apply decorative coatings so that the fabricated item matches the appearance or color coding of adjoining materials.

It is important to coat or paint all cut edges or sanded surfaces of RP to eliminate areas that would invite corrosive attack. Epoxy coatings as well as polyester resin (catalyzed) can be used for this operation.

Remember that nearly any RP shape or profile always has at least two exposed ends where it has been cut off to the desired length after being formed in the continuous process.

For decorative finishes, epoxy and polyester paints are recommended. Acrylic lacquer, vinyls, and oil-base paints can also be used effectively.

For best results, sand the fiber glass surface lightly and follow the paint manufacturer's directions carefully. The following tools and materials may be useful to the person fabricating with structural shapes:

Small hand-held power drill	Fiber glass cloth, mat and tape
Hacksaw	C-clamps
Saber saw	Toluol, acetone, and solvents
Rasps—round and flat	Adhesive cements
Necessary lay-out tools	Mixing cups
Hand sander	Stirring sticks (tongue depressors)
Power-driven sander	Wiping rags
Sandpaper	Mechanical fasteners

Shown below are some typical connections which can be mechanically fastened with self-tapping or standard screws, and with commercial grade rivets or bolts, or bonded with liquid adhesives like epoxy or polyester cement.

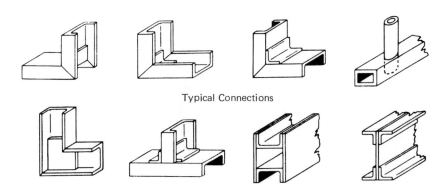

Typical Connections

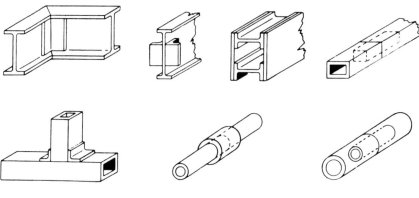

Typical Connections

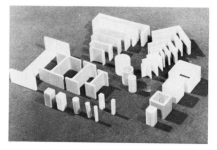

Glass reinforced polyester structural shapes
produced by continuous pultrusion by
Morrison Molded Fiber Glass Company,
Bristol, Va.

Protection fence for high voltage communication equipment.

Structural shapes used as insulating spacers
in this aircraft auxiliary power unit.

Maintenance free ski lift chairs with seat and backrest fabricated from glass reinforced plastic structural shapes by Sierra Engineering, Inc., Reno, Nevada.

Ventilating louver unit with movable blades fabricated from polyester glass reinforced plastic shapes by Imco, Inc., Morristown, N.J.

Screen assembly being towed toward its Lake Ontario installation site. Screens and baffles keep objects larger than six inches in size from entering the underwater intake opening.

A large, water intake screen, fabricated from glass reinforced plastic and concrete. Screen assembly measure 11 feet in height and 56 feet at its widest point. The shield which protects a municipal water plant intake pipe was fabricated by Canbar Industrial Plastics, a division of Canada Barrels and Kegs, Limited.

Framework to support glass reinforced plastic grating for a platform in an electro-plating operation. Platform fabricated from pultruded structural shapes by Fibergrate Corporation. Adjustable screw jacks supporting the platform are also glass reinforced plastic.

A part of a 40 ft. span pipe bridge designed and fabricated by Fibergrate Corporation from FRP pultruded structural shapes. FRP bolts and nuts were used in the assembly eliminating all metallic fasteners.

Stairway fabricated from pultruded structural shapes with glass reinforced plastic grating treads. (Fibergrate Corporation. Dallas, Texas)

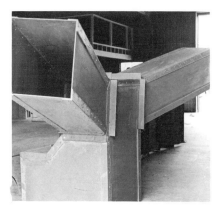

Fume ducts and assemblies fabricated from reinforced-plastic structural shapes and sheet. R. W. Fowler & Associates, Atlantic Beach, Fla.

Washer hood from structural shapes.

Handrail system fabricated from polyester glass reinforced plastic shapes by Imco, Inc., Morristown, N.J.

23

Adhesive Practice

Adhesives, properly selected and used, can provide strong and durable bonds between reinforced plastics and other structural materials. Experience has shown, however, that the strongest joint or connection between reinforced plastics can be made by using mechanical fasteners along with adhesives.

Satisfactory bonds in reinforced-plastic structures are obtained if the joint is properly designed, the mating surfaces are properly prepared, and the recommended types of adhesives are used.

Adhesive joints should be designed to avoid excessive "peel" stresses. Below is an illustration of the four common types of joints and the types of stress that are encountered in each.

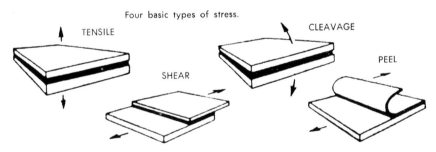

Four basic types of stress.

The compression joint is similar to the tensile joint except that the stress forces are both reversed (compression). Generally speaking, a compression joint is stronger than a tensile joint. In both the stresses are at right angles to the plane of the joint.

In a shear joint, either compression or tensile shear, the joint is subjected to stresses parallel to the plane of the joint. Joints of this type are not quite as strong as the pure compressive-tensile joints, but they are adequate for most structural requirements.

In peel joints and cleavage joints, the bond is subjected to stresses at some intermediate angle, and, as a result, there is a prying or peeling effect at the joint.

An adhesive joint which is weak in peel, as well as a joint undergoing cleavage stresses, should be avoided wherever possible. If peel stresses cannot be avoided in the design, the adhesive joint should always be supplemented with a mechanical fastener or reinforced with additional fiber glass and resin, a technique known as taping.

Two adhesives generally used with reinforced plastics are polyesters and epoxies. If the structure to be bonded is for use in a corrosive environment and the laminate is polyester, the adhesive should be a polyester with corrosion resistance comparable to that of the polyester resin used in the laminate.

Where corrosion resistance is not a consideration, an epoxy adhesive may be more convenient to use. The alleged inconvenience of polyester adhesives is merely the difficulty of measuring the small amount of catalyst required for mixing with the polyester to make the reaction proceed. The inexperienced operator should seek recommendations from the adhesive manufacturer as to the suitability of the adhesive proposed for the specific application.

The remaining element in a good adhesive bond is preparing the mating surfaces and making the adhesive joint. The important steps are summarized:

1. Remove any surface film. Abrade surfaces to be bonded with sandpaper.
2. Remove from surface any dust developed by sanding.
3. Follow manufacturer's directions carefully in preparing adhesive.
4. Spread thin coat of adhesive on both surfaces to be joined.
5. Appropriately clamp assembly to permit adhesive to cure or harden.

Throughout the procedure avoid contamination of any of the surface to be bonded.

With some adhesives there may be modifications of the above general procedure; for example, induction time. Here the directions and recommendations of the manufacturer should apply.

Structural adhesives are inherently more complex materials than their ordinary counterparts, contact and pressure-sensitive adhesives used in day-to-day industrial bonding applications.

Their high strength, chemical resistance, and other desired characteristics are achieved through chemical reaction brought about by catalysts or curing agents (hardeners), or by heating during the curing cycle. Generally speaking, applying a structural adhesive is equivalent to initiating and controlling a chemical reaction. The user must exercise attention to such factors as times, temperatures, and pressures involved, as well as mixing, pot life, and handling requirements.

Appropriate selection, therefore, involves choosing an adhesive system with: the desired strength properties, flexibility, impact resistance. and chemical resistance.

In addition, the system must be compatible with production methods and schedules and with methods of application and cure. Finally, the shelf life, the coverage, and the cleanup requirements must enter into consideration.

Every user of structural adhesives should realize that the ultimate criterion in selection is the complete testing of prototypes bonded together with the adhesive being considered.

Factors such as cleanliness of surfaces, pressures applied during bonding, completeness of cure, as well as the stresses to which the part will be subjected in actual use, are of paramount importance to successful performance. Information supplied by manufacturers may on occasion turn out to be a poor substitute for actual testing of the part. It is important to remember that no manufacturer can be responsible for actions after the material leaves his plant; for example, mixing, application, and temperatures.

The user who needs a high-performance structural adhesive should consult with several reputable manufacturers before making his final selection of material for testing.

24

Safety Practices

A common catalyst used in hand lay-up fabrication techniques is methyl ethyl ketone peroxide (MEKPO). MEKPO is a potentially explosive material and should be carefully handled. Cobalt naphthenate, an accelerator, is used with MEKPO to increase the reactivity of the catalyst system and, accordingly, shorten the gel time. *MEKPO should never be mixed with cobalt naphthenate directly.* A violent reaction or a fire can occur. Most polyester resin suppliers will supply the resin mixed with the desired amount of cobalt naphthenate when requested to do so. Concentration of cobalt naphthenate will vary from about 0.3 to 1.0 percent by weight based on the weight of the resin.

In working with polyester resins in a plant, it may be necessary to purchase small quantities of MEKPO for use. The standard MEKPO is generally supplied in a 60 percent concentration, having been diluted with dimethyl phthalate (to improve catalyst stability). Lesser concentrations are available on special request.

It is good practice to purchase the diluted MEKPO in individually sealed polyethylene (squeeze) bottles, each bottle containing 10 to 15 ml. The details of purchase should be worked out with the source of supply and the purchasing department to eliminate shopping around. Changing MEKPO suppliers for cost reasons is inadvisable since too much can go wrong and the cost is insignificant in relation to the cost of the entire system.

Employees should be permitted to withdraw only one day's supply at a time from storage and to return unused portions of bottles as soon as practicable after completion of the job.

A good way to measure necessary quantities is to use 1 ml. disposable syringes which can be purchased at most drug stores.

Employees should be provided with safety goggles, or shields, and disposable gloves when they are mixing MEKPO with polyester resins.

A thorough understanding of all safety precautions is required for each and every person who uses MEKPO in his work.

All quantities of MEKPO should be stored in their original containers, away from all sources of heat such as steam pipes, radiators, flames, sparks, etc. Containers should be kept closed to avoid contamination. The only satisfactory storage area for MEKPO is one in which the temperature does not exceed 90°F, and, preferably, one where the temperature is maintained at 75°F or less. It is good practice to store MEKPO in a refrigerator if one is readily available.

The importance of preventing overheating is shown in the table below which tabulates half-lives. Half-life is the time required for the decomposition of one-half of the peroxide content originally present in the mixture.

Degrees, F	Half-Life, Hours
180	118
185	91
212	19
239	4
560	less than 15 minutes

For maximum shelf life the MEKPO should be stored at temperatures below 90°F. At this temperature it is regarded as stable.

Storage away from other products is considered best—preferably a separate fireproof building if large quantities are to be stored. In any case, the material should be stored away from strong acids, accelerators, and other easily oxidized compounds. The storage building or area should be unheated, clean, and free of combustible material, and the area around it should be kept free of weeds, trash, etc. "No Smoking" should be posted in the area and unauthorized and untrained personnel should be kept out. Packages should not be left unattended on shipping docks and in areas where they may be exposed to direct sunlight.

From a practical standpoint, one pound of MEKPO will catalyze about 100 pounds of polyester resin. For a small maintenance operation, 10 pounds of MEKPO might represent about a month's supply.

No dispensing, weighing, or transferring of MEKPO should be permitted in the storage area but should be accomplished in a separate location. In all storage or working rooms, there should be an open-top steel drum (20- to 30-gallon size) partially filled with vermiculite or perlite to be used to soak up any spillage. If it is necessary to use the absorbent to soak up a spill, do not use a steel shovel to clean up as it may cause sparking.

If it becomes necessary to dispose of contaminated MEKPO, or vermiculite-soaked MEKPO because of a spill, the methods described below can be used.

A convenient method of disposal is burning in an isolated area. A trench is dug and the liquid distributed along its length. It is ignited with a kerosene rag on the end of a six-foot metal rod. Burning will be rapid but not violent. Local air pollution regulations should be observed in making this choice.

Should the burning method be unacceptable, then the material may be disposed of by hydrolysis, which should be accomplished slowly with constant agitation. The MEKPO should be added slowly to 10 times its weight of 20 percent sodium hydroxide solution. The resulting reaction is mildly exothermic and initially there may be some foaming. The reaction is complete in 24 hours or when no further bubble evolution is observed. At this point the residue solution is no longer hazardous.

The main hazard in the handling of MEKPO is fire. MEKPO burns somewhat sluggishly during the initial period, but when the thermal decomposition point is reached it will burn fiercely, like gasoline. Since MEKPO contains enough active oxygen to support its own combustion, smothering tactics may not always be successful. Copious amounts of water along with foam extinguishers should be available in the event of fire.

MEKPO should not be brought into prolonged contact with the skin, as severe irritation may result. Generally, no harmful effects will be experienced if the area is washed promptly with soap and water. In the case of any severe exposure, it is wise to consult a physician since, in some cases, infection in the surrounding membranes has resulted.

If MEKPO is swallowed, a physician should be promptly consulted. There is no indication that the material has any serious toxic effect; however, it is highly irritating to the mucous membranes. The usual first-aid procedure is to dilute the stomach content with water and induce vomiting.

In utilizing epoxy materials, it is recommended that personnel observe certain safety precautions. When mixing and/or applying epoxy materials, the employee should wear gloves, sleeves, and eye protection equipment. Epoxies will stain clothing and discolor and irritate the skin. Every effort should be made to avoid body contact with epoxy materials.

If the epoxy material comes in contact with the skin, the area should be immediately flushed with copious quantities of water followed by vigorous scrubbing with soap and water. If the material enters the eyes, they should be flushed with copious quantities of water and medical attention should be sought immediately. Never use solvents to try to remove epoxy materials from the skin. Seek medical assistance in cases of extensive exposure or bodily contact.

Epoxy material may cause an allergic reaction in some employees. Experience has shown that some persons are more sensitive to skin reactions from epoxy materials than others. It has also been found that persons apparently immune to skin reactions may become sensitized after undetermined amount of exposure and develop unusual degrees of allergic reactions. On occasion it may become necessary to reassign some workers to other duties.

When an allergic reaction to epoxy material is evident or anticipated, a water-removable protective cream should be used on any exposed body surfaces. Disposable polyethylene gloves should be worn at all times when handling or mixing epoxy materials.

The supervisor should be alert to cases of skin irritation where the employee has been handling fiber glass materials. This type of irritation is different in persistence from that produced by the resins, and will generally subside after bathing.

25

Maintenance Organization

As the use of reinforced plastics continues to advance, many companies are establishing small maintenance organizations to work with this versatile material. This discussion is devoted to suggestions regarding the establishment of a proficient "fiber glass section," or a separate maintenance organization. Local conditions in a plant must prevail in establishing any new or additional organization. These suggestions are possible approaches to the problems and plans should be made in keeping with the local rules and geographical considerations.

Typically, there are more ways than one to do the job.

First, the decision must be made as to whether there is sufficient work to be done with reinforced plastics to justify an organization to work with these materials for 12 months of each year. Experience to date indicates that the best results are achieved with reinforced plastics if a separate organization is set up rather than adding the work to an existing organization and thereby having to train a large number of people.

The type of operation, exposure to the elements, and local plant conditions may limit operations in fiber glass repair to the warm months of the year. Curing a resin system is dependent on temperature, and temperatures below 60°F are not conducive to working with reinforced plastics, since the chemical reaction will not proceed properly at these temperatures.

Assuming there is justification for organizing a section for this type of work, the next consideration might be "How many people?" The new group should be kept to a minimum size unless there are circumstances which dictate that repairs to plant and process equipment must be greatly expedited because of past "deferred maintenance practices."

An example situation is one in which there are no unusual circumstances, and from a personnel or organizational standpoint, there is no backlog of experience in working with reinforced plastics.

In such a situation, the new organization most likely will be working on the "learning curve" for a minimum of 6 to 9 months. The situation may permit or require hiring a supervisor who has worked with these materials, but most likely the supervisor will be an employee who has been with the company for a number of years and has complete familiarity with the plant equipment and processes. For the first few months he will be concerned with learning how to work with reinforced-plastic materials, determining where they can be used, and training the personnel assigned to him.

A "starting organization" should be kept to 3 or 4 people including the supervisor. At the time when 2 or more of the workmen are fairly well trained, additions can be made as desired. An organization of 10 people doing this type of maintenance work approaches the maximum desirable size in a total maintenance organization of 150 to 200 people. When the organization exceeds 5 or 6 people, steps should be taken to have plant engineering department personnel become completely familiar with how to engineer with this material. The timing on the use of the engineering personnel and their integration into the training program will depend on what is to be accomplished by the new maintenance section.

If the objectives are only the repair of existing equipment, then it may be desirable to wait a few months before enlisting engineering's aid and assistance. If the objectives include replacement or major repairs of existing equipment, engineering personnel should be involved in the organization at an early date and given a prominent role. Engineering can relieve the supervisor of a lot of design and administrative considerations and let him work with his people to get a job done well and on schedule.

The next consideration is where to get the people. There are several local organizational situations to be taken into account, such as: no union, one union representing all of the work force in the plant, or two or more separate craft unions in the plant maintenance department.

In the case of a nonunion plant, the new jobs can be filled readily with the best-qualified people available.

Where there is one union representing the whole plant force, there are generally provisions in the labor agreement that must be followed or there is a recognized established procedure that has been used successfully in the past.

The plant that has its maintenance organization represented by two or more separate craft unions (two or more trades) may require careful handling in adding this new work force to avoid a labor problem. Experience shows that each trade may desire representation and the mixing of trades sometimes produces jurisdictional problems. It is imperative that the industrial relations department of the plant become involved at an early date to avoid unnecessary labor problems.

The most efficient approach is to create a new job described in such a way that any person on that job can perform its full range of duties involved in the use of reinforced plastics or maintenance and repair applications. If this is not possible, it should be recognized that the more the work is chopped up among separate job classifications, the less efficient the new organization will be.

Reasonable qualifications should be established for the jobs in the new organization, probably in the form of "job descriptions."

One of the major uses of reinforced plastics in maintenance work has been the repair of piping systems, and people experienced in this work might be considered for the new organization.

Considerable amounts of reinforced plastics have been used in the repair of duct systems or in the building of new hoods, louvers, dampers, and fume collection systems, and people who have worked with this type of equipment may have good potential for the new job.

Glass-reinforced plastics are replacing a number of items formerly made from wood. Many of these types of equipment are custom-made and are not available from a reputable manufacturer; for example, the head box on a paper machine. The experience and abilities of carpenters should therefore be considered.

People experienced in the repair of pumps, impellers, fans, fan housings, agitators, etc., may have potential, as this is another application where reinforced plastics can be used to advantage in the plant maintenance program.

People with experience in painting, the application of industrial coatings, masonry, or "lead burning," are all potential applicants.

Following this discussion are some suggestions on selection of personnel. If the particular plant is not organized or there is no need to consider trade union employees for the new organization, the information on selection of personnel will be the paramount factor in setting up a "fiber glass repair section."

In summary, the man responsible for establishing a new organization would do well to:

1. Decide if there is a need for a separate organization to do maintenance and repair using resin systems reinforced with fiber glass; discuss the plan with the industrial relations people before taking action.

2. Try to establish objectives—or, more specifically, determine what type of equipment is going to be repaired.

3. Estimate the maintenance requirements for the next 6 to 12 months and decide on the number of employees to be trained initially. Prepare a plan as to how large the organization may become and the rate of growth.

4. Plan time for training the organization to perform repairs and maintenance in a workmanlike manner.

5. Plan for utilizing the design services of the engineering department.

6. Decide on the training program that will be needed.

7. Keep the organization small at the start, but do a good job of training. Avoid situations where employees doing work in "fiber glass repair" are responsible to another supervisor.

8. Since one may not have complete freedom of action in establishing the organization, plan to provide the leadership necessary to make it a "going organization," remembering that after a few months the services of the organization will be in demand.

9. Plan the work carefully, starting as "far up the ladder" as possible. In other words, don't spend efforts laying up flat sheet by hand methods when flat sheet, rod, tubing, and equal leg angles can be bought on the market at a total lower cost to the company than the cost of making these items by hand methods.

10. Select a reputable and reliable supplier for all needs. Be sure members of the purchasing department understand these needs and why "buying on price only" can hamper operations and the accomplishment of established objectives.

26

Selection of Personnel

This discussion is directed toward the person responsible for setting up a maintenance organization to do repair work with polyester or epoxy resins reinforced with fiber glass. It offers suggestions on selection of personnel for "manning" the new "fiber glass" organization. Existing union contracts and procedures may restrict freedom of action insofar as the selection of personnel is concerned—except for the supervisory personnel.

SUPERVISOR OF SECTION

A good hunting ground for a section supervisor would be the mechanical and/or maintenance department—persons who have filled the job of assistant master mechanic, or assistant to the master mechanic, or head pipe fitter.

The potential supervisor should be completely familiar with the plant equipment and the plant processes involved, should have actual experience in day-to-day maintenance, and a complete awareness of safety practices.

Experience has shown that persons who have owned and maintained a fiber glass boat may be potential applicants. If the applicant has not demonstrated his ability to work with his hands, his chance of success in this job may be limited.

In addition to all the normal qualifications that one expects in a good supervisor, the applicant should, of course, have an education at least *equivalent* to that of a high school graduate. He should be able to follow written directions; have real desire to further his educational equivalence; his inquisitiveness should extend to the point of his being "hungry for information."

He must expect to do a lot of reading and possibly home study during the early months. He should have a working knowledge of at least several of the principles and fundamentals listed below and be willing and able to further develop his knowledge on all of them.

Corrosive fluids
Strength of materials
Heat transmission
Thermal expansion
Flow of fluids
Pressure and vacuum
Fastening devices
Adhesives
Mixing of materials
Paints and coatings
Structural terms and their meanings
Ability to read drawings and understand specifications

The list above is not intended to be all-inclusive or to qualify or disqualify the potential candidate, but if he exhibits only a "smattering of ignorance" on the above, he may not be the right man. The listing, therefore, may be helpful in the preparation of a job description.

On the assumption that supervisory capabilities have previously been demonstrated, the characteristics or attributes—listed in order of importance—the new supervisor should have are:

1.   Ability to learn and willingness to do outside study.
2.   Willingness and ability to teach others.
3.   Familiarity with plant processes, equipment, and maintenance.
4.   Basic education.

Portions of this book might be used as a check list in interviewing the potential supervisor.

SECTION WORKMEN

The selection of people to man the "fiber glass organization," as previously pointed out, may not allow as much freedom of choice as there is in choosing its supervisor.

The hunting ground, obviously, will be the existing maintenance section.

Experience in some areas of the industry tends to indicate that female employees are generally more proficient working with products made by hand lay-up techniques, particularly if close tolerances on placement of inserts, or thickness of finished laminate, are controlling. This is not to suggest, however, that either sex be favored, but that manual dexterity is important.

It is desirable that the employee selected be genuinely willing to learn. This may not be apparent initially, but perhaps the supervisor can discover and cultivate it by providing an interesting training program.

Assuming that this type of organization is new to the plant, job descriptions should be prepared and used appropriately in selecting employees to fill the new jobs. The preparation of these descriptions is best handled jointly by the supervisor and those responsible for labor relations in the plant.

Some qualifications that may be helpful in preparing a job description are:

1. Ability to read and understand blueprints and drawings.
2. Familiarity with metric system of measurement of volume, length, and weights.
3. Familiarity with Centigrade and Fahrenheit temperature scales and ability to convert one temperature reading to the other.
4. Ability to read and understand technical instructions in connection with special products as supplied by outside manufacturers, particularly when materials that require safe handling are involved.
5. Ability to use properly power-driven hand tools such as sanders, drills, saws, routers, etc.
6. Familiarity with pipe schedules, sizes, and fittings.

In addition, it is assumed that general qualifications on safe practices will be considered. Some of the suggested qualifications for the supervisor may be of use in preparing the job description for the section workmen.

Since a number of resin materials may be classed as skin irritants, the need for wearing protective clothing and practicing good personal hygiene is most important. Experience has also indicated that persons with light complexions tend to be more susceptible to irritation from epoxies and polyesters.

In summary, special care coupled with special effort in the selection of personnel to work with this new construction material, which can be of significant economic benefit to the user, will pay dividends in the years to come.

27

Training of Personnel

The supervisor should be prepared to start immediately on a training program for himself and the people assigned to him. It should be a mixture of theory and practice—on-the-job training. Chapters 17 to 21, which describe how to work with reinforced plastics, can serve as a basis for the training program.

After the information in these sections has been utilized, the supervisor can use spare plant equipment, such as fans, fan housings, pumps, impellers, duct work, and pipe as additional items to expand the training. By careful planning, the supervisor will find that every new job can present some new challenge and fit into the day-to-day advancement of training.

This book is not intended to provide complete information on methods of instruction, but it may be useful to list some of the fundamentals.

1. Limit teaching lectures to a maximum of fifteen minutes at any one time. Explain the basic elements and then get a discussion going with the group.
2. Try to use the coach and pupil technique as much as possible.
3. Try to borrow movies or training films from manufacturers of basic materials.
4. Contact manufacturers for technical literature.
5. Spend at least 50 percent of each training session in demonstrating how something is done properly.
6. Continue to emphasize that good craftmanship is important.
7. In presenting a subject or lesson, there are several basic steps, listed in sequence, to be followed: preparation (by the instructor); explanation (by the instructor); demonstration (by instructor or student); applications (by instructor or student); examination (to check if the student understood what was presented); discussion (initiated by instructor, but continued by the student).

One should remember that an examination can sometimes show more about the abilities of the instructor than the progress of the student.

The supervisor should call on technical representatives from supplying companies to give presentations and demonstrations on their specific products. Many of the companies selling fiber glass piping systems are anxious to make such presentations. The supervisor and the product representative should jointly plan the presentation to avoid repetitious subject material.

The supervisor should plan his instruction sessions for the first part of the morning rather than late afternoon.

Before starting an instruction period or program, he should have the necessary resins, reinforcements, etc., on hand so that the members of his section can actually work with these materials.

Another technique, for later use, is a critique of each job that has been completed. The actual performance of the job should be discussed and items of interest pointed out that may permit a similar job to be accomplished in the future in a more proficient manner.

TRAINING PROGRAM

Since each training situation will be different, the supervisor will have to decide on the particular training needs for the new organization. He should consider the present level of training of the people, their background, experience, and education, and the objectives of the training mission—in other words, what is desired of the group?

The chart below suggests some of the fundamentals that should be covered and proposes a 100-hour program spread over a twelve-month period. The program can easily be escalated up or down at any time by the supervisor. It is desirable that 60 to 70 percent of the program be covered within the first two months after the formation of the organization.

First Two Months Total Hours	Subject	Remaining Ten Months Total Hours
8	Resins and Resin Systems	4
4	Reinforcements	4
4	Safe Practices	4
4	Design of Laminates	
2	Engineering Aspects	4
2	Inspection	4
8	Preparation of Laminates	
8	Repairs with Resin & Fiber Glass	
16	Fabrication	16
2	Training Films	2
2	Outside Suppliers	2
60 hours		40 hours

The development and execution of an interesting training program is strictly the responsibility of the supervisor, and he will find that its accomplishment will thoroughly challenge and test his ingenuity. By accepted teaching standards, 4 hours of preparation are required on the average for one hour of presentation. The above 100-hour program can easily represent 200 to 300 hours of preparatory work by the instructor.

One result that is sure to come out of such a training program is that, if it is properly and adequately presented, the instructor may find he has learned much more than his pupils.

The supervisor, once selected, should have at least one month to get his "house in order" before receiving the "students." His time can be well spent in ordering materials, planning the training program, planning the shop area or working area, ordering tools and perhaps doing a littled selected reading to broaden his perspective and add to his background.

Please refer to the appendix for additional information on training aids.

28

A Maintenance Shop for Reinforced Plastics

When preparing an area for working with reinforced plastics, one of the first considerations should be to eliminate all known hazards. Following good industrial practices—now a routine matter in most well-run organizations— provide an area where no problems with either the health of the personnel concerned or the possibility of fire will be experienced.

Assuming that for the most part the resin systems used will be polyester, the monomer usually present will be styrene. OSHA standard (1973) for eight-hour exposure to styrene vapors is 100 parts per million. Studies have shown that acute toxic action resulting from excess inhalation of styrene vapors arise when the concentration exceeds 400 parts per million. At this concentration, styrene vapors are disagreeable to most people.

The molding shop should have a temperature-controlled ventilating system which will provide a minimum of 4 air changes per hour. If any spraying operations are to be carried out, a special booth should be provided.

As discussed previously, the resin and monomer quantities over and above the day's requirements should be stored in a separate building away from the fabrication shop. Proper storage should also be provided for the catalysts. Separate storage should be provided for solvents, but they can be stored in the same building as the resins. "No Smoking" should be posted for all storage areas. The glass-reinforcement materials can be stored in the fabricating shop but they should be protected from dust, dirt, and dampness. Using reinforcements which have been subjected to dampness or extreme humidity results in poor quality laminates.

Depending on the requirements of the plant, there may well be a need for an area for fabricating into finished form the items made from reinforced plastics. In such an area, parts made by the lay-up shop would be used with structural shapes purchased from suppliers. Since operations like sawing, drilling, sanding, and adhesive bonding would be carried on in the fabricating area, the immediate area could not be used for hand lay-up operations.

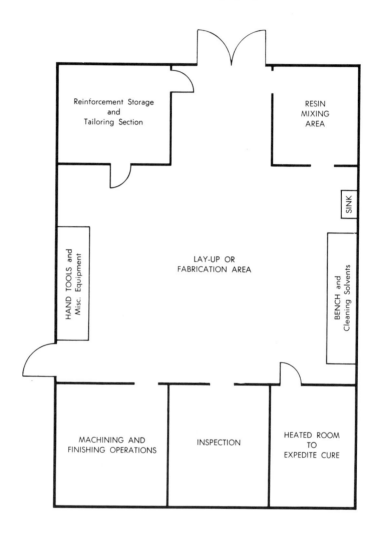

29

Cost Records

The small fabrication or maintenance shop may find it desirable and/or necessary to maintain simple cost records. Such records can serve several purposes:

1. Provide a measure of control and economic productivity for the supervisor.
2. Serve as a basis for estimating future jobs.
3. Assist in deciding whether to do the job in-house or to purchase on the outside.
4. Provide a basis for judging when to laminate standard shapes (flats, sheets, angles, tubes, etc.) or to purchase on the outside.

As experience is gained in conducting a small fabrication operation, the economic utility of purchasing shapes from suppliers who produce them by continuous machine methods, rather than being totally self-sufficient and hand fabricating every piece, will become more and more apparent.

Proposed simplified cost records follow.

OUR OWN SHOP

Reinforced Plastics

Job No. _____ Date Assigned _____ Date Completed _____

Name or Description _____

Reinforcement Used

Pounds of Mat _____X$_____per pound = $_____

Pounds of Cloth _____X$_____per pound = $_____

Pounds of Roving	_____	X$_____ per pound = $_____
Pounds of Surfacing Mat	_____	X$_____ per pound = $_____
Pounds of Cloth	_____	X$_____ per pound = $_____
Pounds of _____	_____	X$_____ per pound = $_____
			(A)

Resins

Pounds of Resin	_____	X$_____ per pound
	_____	X$_____ per pound
	_____	X$_____ per pound
	_____	X$_____ per pound $_____

Total Pounds _____ (B)

Add 15% of Resin Cost for Catalysts (C) $_____

(A + B + C) Total Raw Material Cost (D) $_____

Raw Material Cost per Pound = D ÷ Total Pounds $_____

Labor Costs

_____ Hours X Rate = (E) $_____

Miscellaneous Materials   _____

Tooling Costs (F) $_____

Total Material and Labor (G) $_____

Net Weight of Unit Constructed _____ pounds (H)   _____

Control Factors

$$\text{Conversion Efficiency} = \frac{\text{Net Weight of Finished Item}}{\text{Total Weight of Raw Materials}} \times 100 = \underline{\qquad} \%$$

$$\text{Wastage Factor} = 100 - \text{Conversion Efficiency} = \underline{\qquad} \%$$

$$\text{Net Cost of Finished Item} = \frac{\text{Total Cost (G)}}{\text{Wt. of Item (H)}} = \$\underline{\qquad} \text{ PER LB.}$$

$$\text{Conversion Cost} = \frac{\text{Total Labor \& Tooling Costs (E + F)}}{\text{Net Weight of Finished Unit (H)}} = \$\underline{\qquad} \text{ PER LB.}$$

$$\text{Material Cost} = \frac{\text{Total Cost of Resins and Reinforcement (D)}}{\text{Net Weight of Finished Unit (H)}} = \$\underline{\qquad} \text{ PER LB.}$$

Material Cost/lb. plus Conversion Cost/lb. should = Net Cost per Pound of Unit.

Net Cost of Finished Item = $_____ PER LB.

EXAMPLE OF COST RECORD (SUMMARY)

Assuming that a simple laminate is to be built which will use 80 pounds of glass at an average cost of 60 cents/lb., and 60 pounds of resin at 30 cents/ lb., the finished weight of the unit is 120 pounds. The labor for completing the laminate is 16 hours at $3.20 per hour. This includes fabrication, cleanup, cure, trimming, and supervision.

Reinforcement Used

80 pounds at $.60 _____ $48.00

Resins

60 pounds at $.30 _____ 18.00

Catalysts, etc., at 15% _____ 2.70

    Total Raw Material Cost $68.70

Raw Material Cost per Pound = $68.70 ÷ 140 = $.49 per Pound

Labor Costs

16 hours at $3.20 _____ = $51.20

Net Weight of Unit Constructed _____ =   120 lbs.

$\text{Conversion Efficiency} = \dfrac{120}{140} \times 100$ _____ =   85.6%

Wastage Factor = 100 − 85.6 _____ =   14.4%

$\text{Conversion Cost} = \dfrac{\$51.20}{120}$ _____ = 42.7 cents per lb.

$\text{Material Cost} = \dfrac{\$68.70}{120}$ _____ = 57.3 cents per lb.

$\text{Net Cost of Finished Item} = \dfrac{\$119.90}{120}$ _____ = $1.00 per lb.

    The cost figures shown in the above example will vary with respect to location and experience. In evaluating the example, the actual work involved could represent about 140 square feet of 1/8″-thick laminate, or it could be one sheet 4′ × 8′, about ½″ thick. The example is presented only to illustrate a simple method of analyzing costs.

    An alternate method would be to use this same system on a monthly basis. Here the supervisor would account for the reinforcements and resins used during the month, together with his labor and total pounds of finished product. This method will provide, over a long period of time, a basis for judging the success of the operation.

    Following are the approximate commercial prices for reinforced plastics:

Flat sheets	Less than $1.00 per pound
Simple structures	$1.25 to $2.50 per pound
Structural shapes	$1.00 to $1.50 per pound

Tubing	$1.50 or less per pound
Pipe (pressure)	$1.50 to $3.00 per pound

Cost records can be used as a guide in deciding whether to fabricate all items needed, or to start farther ahead by purchasing standard products on the outside and fabricating them into needed structures. The appropriate decision will not only cut down on the overall backlog of jobs, but will improve the effectiveness and productivity of the entire maintenance department.

APPENDIXES

Measurement Chart
For Adding % Weight Catalyst

Volumetrically in Milliliters or Fluid Ounces

Amount of Resin	For 1/2%		For 3/4%		For 1%		For 1-1/2%		For 2%	
	Ml.	Oz.	Ml.	Oz.	Ml.	Oz.	Ml.	Oz.	Ml.	Oz.
1 Pint	2-1/4	1/16	3-1/2	1/8	4-1/2	1/8	7	1/4	9	3/8
1 Quart	4-1/2	1/8	7	1/4	9	1/3	14	1/2	18	5/8
#10 Can	15	1/2	23	3/4	30	1	45	1-5/8	60	2
1 Gallon	19	5/8	28	1	37	1-1/4	56	2	74	2-1/2
5-Quart Oil Can	23	3/4	35	1-1/8	46	1-5/8	70	2-1/2	92	3-1/4
5-Gallon Pail	97	3-1/4	140	4-3/4	185	6-1/2	280	9-3/4	370	13

Example: If formula calls for 1/2% by weight of MEKPO, add 2-1/4 Ml. or 1/16 oz. by volume to one pint of resin.

All amounts are rounded off to nearest fraction.

Polyester resin is based on 9.2 lbs. per gallon. Specific gravity of MEKPO = 1.13.

Weight Versus Volume Measurements for MEKPO (60%) for Resin at 9.2 lbs./gal.

Weight	% By Weight of Typical Resin	Fluid	Milliliters	Standard Teaspoons	Drops from Standard Eye Dropper
1 Oz.	2.7	0.9	27	6	700

Pressure Conversion Table

Pounds Per Square Inch	Pressure in Pounds Per Square Inch or Feet of Water to Be Converted	Feet of Water Head
0.43	1	2.31
0.87	2	4.62
1.30	3	6.93
1.73	4	9.24
2.17	5	11.54
2.60	6	13.85
3.03	7	16.16
3.46	8	18.47
3.90	9	20.78
4.33	10	23.09
8.66	20	46.18
12.99	30	69.27
17.32	40	92.36
21.65	50	115.45
25.98	60	138.54
30.31	70	161.63
34.65	80	184.72
38.98	90	207.81
43.31	100	230.90
54.14	125	288.67
64.96	150	346.41
86.62	200	461.88
129.93	300	692.82
173.24	400	923.76
216.55	500	1154.70

Note: Based on one pound of pressure per square inch equals 2.309 feet of water at 62° F.

Temperature Conversion Table
Centigrade and Fahrenheit

$°C$	*Temp. of $C°$ or $F°$ to be Converted*	$°F$
−73.33	−100	−148.0
−67.77	− 90	−130.0
−62.21	− 80	−112.0
−56.66	− 70	− 94.0
−51.11	− 60	− 76.0
−45.55	− 50	− 58.0
−40.00	− 40	− 40.0
−34.44	− 30	− 22.0
−28.88	− 20	− 4.0
−23.33	− 10	14.0
−20.55	− 5	23.0
−17.77	0	32.0
−15.00	5	41.0
−12.22	10	50.0
− 9.44	15	59.0
− 6.66	20	68.0
− 3.89	25	77.0
− 1.11	30	86.0
1.67	35	95.0
4.44	40	104.0
7.22	45	113.0
10.00	50	122.0
12.78	55	131.0
15.55	60	140.0
18.33	65	149.0
21.11	70	158.0
23.88	75	167.0
26.66	80	176.0
29.44	85	185.0
32.22	90	194.0
35.00	95	203.0
37.77	100	212.0
43.33	110	230.0
48.88	120	248.0
60.00	140	284.0
65.55	150	302.0
93.33	200	392.0
98.88	210	410.0
104.44	220	428.0
110.00	230	446.0
121.11	250	482.0
148.88	300	572.0

Polyester Resin Producers

Below is a partial list of resin producers. Technical information may be obtained from the manufacturer on request.

ACME Resin Company, Forest Park, Il., 60130
Alpha Chemical & Plastics Corp., Newark, N.J. 07105
American Cyanamid Co., Wayne, N.J. 07470
Ashland Chemical Co., Columbus, Oh. 53216
Atlas Chemicals Div., ICI America, Inc., Wilmington, De. 19899
Cargill, Inc., Lynwood, Ca. 90262
Cook Paint and Varnish Co., Kansas City, Mo. 64141
Diamond Shamrock Chemical Co., Cleveland, Oh. 44115
Durez Div., Hooker Chemical Corp., N. Tonawanda, N.Y. 14120
Freeman Chemical Corp., Div. H.H. Robertson Co., Port Washington, Wi. 53074
Inmont Corp., Clifton, N.J. 07015
Interplastic Corp., Minneapolis, Mn. 55413
Koppers Company, Inc., Pittsburgh, Pa. 15219
Marco Chemical Div., W. R. Grace & Co., Linden, N.J. 07036
Mol-Rez Div., Whittaker Corp., Minneapolis, Mn. 55418
Owens-Corning Fiberglas Corp., Toledo, Oh. 43601
PPG Industries, Inc., Pittsburgh, Pa. 15222
Reichhold Chemicals, Inc., White Plains, N.Y. 10602
Rohm & Hass Co., Philadelphia, Pa. 19105
Schenectady Chemicals, Inc., Schenectady, N.Y. 12301
Sherwin-Williams Co., Cleveland, Oh. 44101
Silmar Div., Vistron Corp., Hawthorne, Ca. 90250
Whittaker Coatings & Chemicals, Lenoir Coatings and Resins Div., Lenoir,
    N.C. 28645

Training Aids

A versatile color-slide presentation on reinforced plastics/composites has been prepared by the Society of the Plastics Industry, Inc., 250 Park Avenue, New York 10017. This 30-minute presentation, "RP/C: The Shape of Tomorrow— Today," is complete with 87 projection slides and script. It presents an overview of the tremendous range of uses, the advantages, and the future potential of reinforced plastics/composites (RP/C). It also covers materials used and production processes. The slide carousel, included in the purchase price, can accept an additional number of slides so that the presentation can be tailored to emphasize selected areas of interest.

The following .16 mm.-sound color films are available from Goldworthy Engineering, Inc., 2917 Lomita Blvd., Torrance, Ca., 90505, on short-term loan:

The Glastruder describes pultrusion as a generic process; features the Glastruder pultrusion machine with precise definition of machine components; describes Glastrusions patented pultrusion process. 15 minutes.

Filament-Winding Systems defines filament winding; features a variety of filament-winder machine types: box winder, lathe, race track, and polar, with a variety of control systems including digital control. 24 minutes.

Box Winder features one machine, a box winder, proving that objects other than bodies of revolution can be filament-wound. 25 minutes.

Arrangements can be made with Mr. Gerald D. Shook, Consultant, 8 Wyoming Drive, Huntington Station, N.Y. 11746 for showing films on automatic spray-up, automatic centrifugal casting, and compression molding.

Other excellent sources of training-aid material are: Owens-Corning Fiberglas Corp., Toledo, Oh. 43601, and PPG Industries, Pittsburgh, Pa. 15222.

Metric Conversion Tables

The unit of length in the metric system, the meter, is intended to be 1 ten-millionth part of the distance from the equator to the pole. The original meter is a platinum bar made in 1799 and kept in the archives of the French Republic. Refinements in the measuring of the earth's surface since then have shown that only a very slight discrepancy exists between the actual and the intended length.

The meter is divided into 10 equal parts, decimeters; each of these into 10 centimeters, and each centimeter into 10 millimeters. A millimeter, therefore, is 1 thousandth part of a meter.

Simultaneously with the meter bar a platinum weight was constructed, as nearly as possible equal to the mass of a cube of pure water at 4°C, and whose side is one decimeter. The weight is called the kilogram and equals 1,000 units or grams.

The volume of one kilogram of pure water at the temperature of maximum density (4°C.) and under a pressure of 76 cm. of mercury is called the liter and is the unit of capacity in the metric system.

The metric units of length, mass, and capacity are subdivided decimally, the Latin prefixes of deci, centi, and milli being used to indicate the order of the divisions, while the Greek prefixes, deka, hecto, kilo, and myria, are used to indicate the order of multiplication of the units by 10.

METRIC SYSTEM

Units

Length—Meter   **Mass—Gram**   **Capacity—Liter**

for pure water at 4°C. (39.2°F.)
1 cubic decimeter or 1 liter = 1 kilogram

$$1000 \text{ Milli} \begin{Bmatrix} meters \text{ (mm)} \\ grams \text{ (mg)} \\ liters \text{ (ml)} \end{Bmatrix} = 100 \text{ Centi} \begin{Bmatrix} meters \text{ (cm)} \\ grams \text{ (cg)} \\ liters \text{ (cl)} \end{Bmatrix} = 10 \text{ Deci} \begin{Bmatrix} meters \text{ (dm)} \\ grams \text{ (dg)} \\ liters \text{ (dl)} \end{Bmatrix} = 1 \begin{matrix} meter \\ gram \\ liter \end{matrix}$$

$$1000 \begin{Bmatrix} meters \\ grams \\ liters \end{Bmatrix} = 100 \text{ Deka} \begin{Bmatrix} meters \text{ (dkm)} \\ grams \text{ (dkg)} \\ liters \text{ (dkl)} \end{Bmatrix} = 10 \text{ Hecto} \begin{Bmatrix} meters \text{ (hm)} \\ grams \text{ (hg)} \\ liters \text{ (hl)} \end{Bmatrix} = 1 \text{ Kilo} \begin{Bmatrix} meter \text{ (km)} \\ gram \text{ (kg)} \\ liter \text{ (kl)} \end{Bmatrix}$$

U.S. to Metric

Metric to U.S.

Linear

1 inch = .0254001 meters.
1 foot = .304801 meters
1 yard = .914402 meters
1 mile = 1609.35 meters
       = 1.60935 kilometers

1 meter = 39.3700 inches
1 meter = 3.28083 feet
1 meter = 1.09361 yards
1 kilometer = .62137 miles

Square

1 sq. inch = 6.452 sq. centimeters
1 sq. foot = 9.290 sq. decimeters
1 sq. yard = .836 sq. meters

1 sq. centimeter = .1550 sq. inches
1 sq. meter = 10.7640 sq. feet
1 sq. meter = 1.196 sq. yards

Cubic

1 cu. inch = 16.387 cu. centimeters
1 cu. foot = .02832 cu. meters
1 cu. yard = .765 cu. meters

1 cu. centimeter = .0610 cu. inches
1 cu. meter = 35.314 cu. feet
1 cu. meter = 1.308 cu. yards

Weight

1 grain = 64.7989 milligrams
1 avoirdupois ounce = 28.3495 grams
1 troy ounce = 31.10348 grams
1 avoirdupois pound = .45359 kilograms

1 milligram = .01543 grains
1 kilogram = 15432.36 grains
1 hectogram = 3.5274 avoirdupois ounces
1 kilogram = 2.20462 avoirdupois pounds

Capacity

1 fluid drachm = 3.70 cu. centimeters
1 fluid ounce = 29.57 milliliters
1 quart = .94636 liters
1 gallon = 3.78544 liters

1 milliliter = .27 fluid drachms
1 centiliter = .338 fluid ounces
1 liter = 1.0567 quarts
1 hectoliter = 26.417 gallons

Pressure

1 pound per square inch = .0703
    kilograms per square centimeter

1 kilogram per square centimeter = 14.22
    pounds per square inch

Metric Equivalents

	Approximate		*Exact*
1 centimeter	0.39	inch	0.3937
1 cubic centimeter	0.061	cubic inch	0.0610
1 cubic foot	0.028	cubic meter	0.0283
1 cubic inch	16.	cubic centimeters	16.39
1 cubic meter	35.	cubic feet	35.31
1 cubic meter	1.3	cubic yards	1.308
1 cubic yard	0.76	cubic meter	0.7646
1 foot	30.	centimeters	30.48
1 gallon (U.S.)	3.8	liters	3.785
1 gallon (Imperial)	4.5	liters	4.546
1 grain	0.065	gram	0.0648
1 gram	15.	grains	15.43
1 inch	25.	millimeters	25.40
1 kilogram (kilo)	2.2	pounds	2.205
1 liter	1.1	quarts (liquid) (U.S.)	1.057
1 liter	0.88	quart (liquid) (Imperial)	0.8799
1 meter	3.3	feet	3.281
1 mile	1.6	kilometers	1.609
1 millimeter	0.039	inch	0.0394
1 ounce (avoirdupois)	28.	grams	28.35
1 ounce (Troy)	31.	grams	31.10
1 pint (liquid)	0.47	liter	0.4732
1 pound	0.45	kilogram	0.4536
1 quart (liquid)	0.95	liter	0.9463
1 square centimeter	0.15	square inch	0.1550
1 square foot	0.093	square meter	0.0929
1 square inch	6.5	square centimeters	6.452

	Approximate		*Exact*
1 square meter	1.2	square yards	1.196
1 square meter	11.	square feet	10.76
1 square yard	0.84	square meter	0.8361
1 yard	0.91	meter	0.9144

Selected Reading

Crosby, E. G., and Kochis, S. N. *Practical Guide to Plastics Applications.* Boston: Cahners Books, 1972.

deDani, A. *Glass Fibre Reinforced Plastics.* New York: Interscience Publishers, 1960.

DuBois, J. H. *Plastics History U.S.A.* Boston: Cahners Books, 1972.

Duffin, D. J. *Laminated Plastics.* New York: Reinhold Publishing Company, 1966.

Gibbs & Cox, Inc. *Marine Design Manual.* New York: McGraw-Hill, Inc., 1960.

Lawrence, J. R. *Polyester Resins.* New York: Reinhold Publishing Company, 1960.

Lee, H., and Neville, K. *Epoxy Resins Applications and Technology.* New York: McGraw-Hill, Inc., 1957.

Mallinson, J. H. *Chemical Plant Design with Reinforced Plastics.* New York: McGraw-Hill, Inc., 1969.

——. *Modern Plastics Encyclopedia.* New York: McGraw-Hill, Inc., 1973.

Mohr, J. G.; Oleesky, S. S.; Shook, G. D.; and Meyer, L. S. *Technology and Engineering of Reinforced Plastics/Composites An SPI Handbook.* New York: Von Nostrand Reinhold Company, 1973.

Oleesky, S. S., and Mohr, J. G. *Handbook of Reinforced Plastics of the SPI.* New York: Reinhold Publishing Company, 1964.

Rosato, Dominick V., and Grove, C. S., Jr. *Filament Winding.* New York: Interscience Publisners, 1964.

Rosato, Dominick V.; Fallon, W. K.; and Rosato, Donald V. *Markets for Plastics.* New York: Van Nostrand Reinhold Company, 1969.

Simonds, H. R., and Church, J. M. *Concise Guide to Plastics.* New York: Reinhold Publishing Company, 1963.

Skeist, I. *Epoxy Resins.* New York: Reinhold Publishing Company, 1958.

——. *Handbook of Adhesives.* New York: Reinhold Publishing Company, 1962.

Index